Governors State University
Library
Hours:
Monday thru Thursday 8:30 to 10:30
Friday and Saturday 8:30 to 5:00
Sunday 1:00 to 5:00 (Fall and Winter Trimester Only)

DEMCO

DATA-DRIVEN BLOCK CIPHERS FOR FAST TELECOMMUNICATION SYSTEMS

OTHER TELECOMMUNICATIONS BOOKS FROM AUERBACH

Architecting the Telecommunication Evolution: Toward Converged Network Services
Vijay K. Gurbani and Xian-He Sun
ISBN: 0-8493-9567-4

Business Strategies for the Next-Generation Network
Nigel Seel
ISBN: 0-8493-8035-9

Chaos Applications in Telecommunications
Peter Stavroulakis
ISBN: 0-8493-3832-8

Context-Aware Pervasive Systems: Architectures for a New Breed of Applications
Seng Loke
ISBN: 0-8493-7255-0

Fundamentals of DSL Technology
Philip Golden, Herve Dedieu, Krista S Jacobsen
ISBN: 0-8493-1913-7

Introduction to Mobile Communications: Technology,, Services, Markets
Tony Wakefield, Dave McNally, David Bowler, Alan Mayne
ISBN: 1-4200-4653-5

IP Multimedia Subsystem: Service Infrastructure to Converge NGN, 3G and the Internet
Rebecca Copeland
ISBN: 0-8493-9250-0

MPLS for Metropolitan Area Networks
Nam-Kee Tan
ISBN: 0-8493-2212-X

Performance Modeling and Analysis of Bluetooth Networks: Polling, Scheduling, and Traffic Control
Jelena Misic and Vojislav B Misic
ISBN: 0-8493-3157-9

A Practical Guide to Content Delivery Networks
Gilbert Held
ISBN: 0-8493-3649-X

Resource, Mobility, and Security Management in Wireless Networks and Mobile Communications
Yan Zhang, Honglin Hu, and Masayuki Fujise
ISBN: 0-8493-8036-7

Security in Distributed, Grid, Mobile, and Pervasive Computing
Yang Xiao
ISBN: 0-8493-7921-0

TCP Performance over UMTS-HSDPA Systems
Mohamad Assaad and Djamal Zeghlache
ISBN: 0-8493-6838-3

Testing Integrated QoS of VoIP: Packets to Perceptual Voice Quality
Vlatko Lipovac
ISBN: 0-8493-3521-3

The Handbook of Mobile Middleware
Paolo Bellavista and Antonio Corradi
ISBN: 0-8493-3833-6

Traffic Management in IP-Based Communications
Trinh Anh Tuan
ISBN: 0-8493-9577-1

Understanding Broadband over Power Line
Gilbert Held
ISBN: 0-8493-9846-0

Understanding IPTV
Gilbert Held
ISBN: 0-8493-7415-4

WiMAX: A Wireless Technology Revolution
G.S.V. Radha Krishna Rao, G. Radhamani
ISBN: 0-8493-7059-0

WiMAX: Taking Wireless to the MAX
Deepak Pareek
ISBN: 0-8493-7186-4

Wireless Mesh Networking: Architectures, Protocols and Standards
Yan Zhang, Jijun Luo and Honglin Hu
ISBN: 0-8493-7399-9

Wireless Mesh Networks
Gilbert Held
ISBN: 0-8493-2960-4

AUERBACH PUBLICATIONS

www.auerbach-publications.com
To Order Call: 1-800-272-7737 • Fax: 1-800-374-3401
E-mail: orders@crcpress.com

DATA-DRIVEN BLOCK CIPHERS FOR FAST TELECOMMUNICATION SYSTEMS

GOVERNORS STATE UNIVERSITY
UNIVERSITY PARK
IL 60466

Nikolay A. Moldovyan
Alexander A. Moldovyan

Auerbach Publications
Taylor & Francis Group
New York London

CRC Press is an imprint of the
Taylor & Francis Group, an **informa** business

QA
76.9
.A25
M664
2007

Auerbach Publications
Taylor & Francis Group
6000 Broken Sound Parkway NW, Suite 300
Boca Raton, FL 33487-2742

© 2008 by Taylor & Francis Group, LLC
Auerbach is an imprint of Taylor & Francis Group, an Informa business

No claim to original U.S. Government works
Printed in the United States of America on acid-free paper
10 9 8 7 6 5 4 3 2 1

International Standard Book Number-13: 978-1-4200-5411-8 (Hardcover)

This book contains information obtained from authentic and highly regarded sources. Reprinted material is quoted with permission, and sources are indicated. A wide variety of references are listed. Reasonable efforts have been made to publish reliable data and information, but the author and the publisher cannot assume responsibility for the validity of all materials or for the consequences of their use.

No part of this book may be reprinted, reproduced, transmitted, or utilized in any form by any electronic, mechanical, or other means, now known or hereafter invented, including photocopying, microfilming, and recording, or in any information storage or retrieval system, without written permission from the publishers.

For permission to photocopy or use material electronically from this work, please access www.copyright.com (http://www.copyright.com/) or contact the Copyright Clearance Center, Inc. (CCC) 222 Rosewood Drive, Danvers, MA 01923, 978-750-8400. CCC is a not-for-profit organization that provides licenses and registration for a variety of users. For organizations that have been granted a photocopy license by the CCC, a separate system of payment has been arranged.

Trademark Notice: Product or corporate names may be trademarks or registered trademarks, and are used only for identification and explanation without intent to infringe.

Library of Congress Cataloging-in-Publication Data

Moldovyan, Nick.
Data-driven block ciphers for fast telecommunication systems / Nikolai Moldovyan, Alexander A. Moldovyan.
p. cm.
Includes bibliographical references and index.
ISBN 978-1-4200-5411-8 (hardback : alk. paper) 1. Computer security. 2. Ciphers. 3. Cryptography. I. Moldovyan, Alex. II. Title.

QA76.9.A25M664 2007
005.8--dc22 2007031200

Visit the Taylor & Francis Web site at
http://www.taylorandfrancis.com

and the Auerbach Web site at
http://www.auerbach-publications.com

Dedicated to the memory of our father,
Andrey Alexander Moldovyan(u),
who always supported us in our creative affairs.

Contents

Preface

Among different directions of applied cryptography, the cipher design based on the data-driven operations (DDOs) is, by comparison, poorly represented in the published and available literature. The RC5 and RC6 block ciphers based on data-dependent rotations (DDRs) are well known, but many other DDO-based ciphers are known only to a small number of readers. The first application of the DDOs as a main cryptographic primitive relates to RC5 [56] and encryption algorithms based on data-dependent subkey selection [32]. The DDO are implemented with so-called controlled operation boxes (COBs), the most known type of which is represented by permutation networks (PNs). The PNs have been well studied in the field of telephone switching systems and parallel computations (1950–1980), see for example [4, 11, 68]. However, the first application of PNs in cryptalgorithm design relates to 1992 [53], where PNs have been applied to perform key-dependent bit permutations. Only about ten years later were the PNs used to perform data-dependent permutations. Dependence of the bit permutations on input data has imparted a new quality to PNs as a cryptographic primitives and solved some principal problems connected with key-dependent permutations. New applications of PNs have been requested to investigate new properties of PNs. In this area differential and linear characteristics of PNs have been studied and the notions of the PN order and switchable PNs have been introduced.

The next step of the development of the DDO-based cipher design relates to generalization of PNs connected with introducing the controlled substitution–permutation networks (CSPNs) constructed using the minimum size controlled elements. It has been shown that ciphers based on the PNs and CSPNs, implementing DDOs of different types, provide fast encryption and low implementation cost in hardware.

One can currently read papers devoted to different particular items on DDOs as a cryptographic primitive, however there is no published work introducing the DDO-based design as an individual section of the applied cryptography. Because the DDO-based ciphers represent an essential interest for application in the fast telecommunication systems and mobile networks, we have found that summarizing the known results in one book to be both reasonable and important.

In this book we present the most interesting results on the DDO boxes topologies, classification of the controlled elements, DDO-based ciphers, and fast software-oriented encryption algorithms using the data-driven subkey section as a main primitive. Prior to this book a major portion of these results have been available only in Russian literature.

Acknowledgments

We are grateful to the editorial board of *The Computer Science Journal of Moldova* for permission to use the material previously published in our papers. These include: 1994, volume 2, pp. 269–282; 2000, volume 8, pp. 270–283; 2003, volume 11, pp. 292–315; 2005, volume. 13, pp. 84–109; 2005, volume 13, pp. 268–291.

We also thank the co-authors of our papers and the authors of other papers, reports, and books cited for their contributions to the development of data-driven block ciphers, the subject of this book.

Introduction

Block data encryption is the most widely used cryptographic transformation in computer and telecommunication systems. The security of ciphers should be based on the secrecy of a small portion of information called a *key*. All other details of cryptosystems are considered as known elements. Moreover, ciphers should be secure against known and chosen text attacks as well as against different side channel attacks. If we mention encryption algorithm, then first of all we have in mind security requirements. Substitution-permutation and Feistel's networks represent conventional designs of block ciphers. Controlled operations have been unsuccessfully tried over a long term as the main cryptographic primitive in cipher design. Recently, they have been proposed to perform variable transformation, i.e., data-dependent (DD) operations (DDOs). This has given birth to a new direction in fast cryptography oriented to cheap hardware implementation. Such application of the controlled operations lead to designing the data-driven ciphers providing significant reduction of hardware implementation costs to relatively conventional designs. The implementation efficacy estimated using different comparison models increases by a factor up to 10 against AES candidates and other conventional ciphers. The data-driven ciphers are very attractive while embedding security mechanisms in constrained environments. Therefore, the data-driven ciphers are extremely interesting for usage to support the successful solution of security problems while deploying ad hoc and sensor networks.

Histories, different types, and topologies of the DDO boxes are considered in this book. Detailed attention is paid to design, properties, and application of the DD permutations (DDPs). Several DDP-based ciphers are considered concerning security and implementation items. Topologies of the controlled substitution-permutation networks (CSPNs) that are built up using small-size controlled elements (CEs) as standard building block are considered. The notion of the order of the CSPN is explained. The classification of CEs corresponding to different types is described. Reversible CSPNs representing a new cryptographic primitive and their designs are discussed. Application of the CSPNs as DDOs is illustrated by examples of several ciphers with 64- and 128-bit data block. A feature of the ciphers is the use of very simple key scheduling providing high performance in case of frequent change of keys.

One of the book's chapters is devoted to another novel primitive—switchable data-driven operations implemented with CSPNs. Different designs of the switchable data-driven operation and its implementation in fast block cipher syntheses are proposed.

The presented implementation comparison of the DDO-based ciphers with other known block encryption systems shows that the first ones are significantly more efficient for embedding in hardware. Regarding software implementation, the DDP-based algorithms are very attractive due to the research on the justification of embedding a controlled bit permutation instruction in the set of standard commands of the general purpose microprocessor. To make the embedding of such instruction more attractive for the CPU manufacture, the book proposes a universal architecture of the controlled bit permutation instruction, which allows performing both the data-driven permutations and the arbitrary prescribed fixed permutation in one cycle.

The book considers, also, the fast software suitable ciphers based on a data-dependent subkey selection, which can be efficiently implemented using the standard instructions of the currently available processors.

About the Authors

Nikolay A. Moldovyan, Ph.D., is an honored inventor of the Russian Federation (2002), Prof. Doctor (2001), a chief researcher with the Specialized Center of Program Systems "SPECTR," and a professor with the Saint Petersburg State Electrotechnical University. His research interests include computer security and cryptography. He has authored or co-authored more than 60 patents and 250 scientific articles, books, and reports. He received his Ph.D. from the Academy of Sciences of Moldova (1981). He can be contacted at: nmold@cobra.ru.

Alexander A. Moldovyan, Ph.D., is a Prof. Doctor (2005), a director with the Specialized Center of Program Systems "SPECTR," and a professor with the State University for Waterway Communications (Saint Petersburg, Russia). His research interests include information assurance, computer security, and applied cryptography. He has authored or co-authored more than 45 patents and 180 scientific articles, books, and reports. He received his Ph.D. from the Saint Petersburg Electrical Engineering University (1996). He can be contacted at: ma@cobra.ru.

Abbreviations

AES	advanced encryption standard
ASIC	application specific integrated circuit
BF	Boolean function
BPI	bit permutation instruction
CBC	cipher block chaining
CE	controlled element
CFB	cipher-feedback
CLB	configurable logic block
COS	controlled operational substitution
CP	controlled permutation
CPB	controlled permutation box
CPI	controlled permutation instruction
CPU	central processing unit (processor)
CSPN	controlled substitution-permutation network
CTPO	controlled two-place operation
DC	differential characteristic
DCA	differential cryptanalysis
DDO	data-dependent operation (data-driven operation)
DDP	data-dependent permutation (data-driven permutation)
DDR	data-dependent rotation
DDSS	data-dependent subkey selection
DFA	differential fault analysis
DFF	D Flip-Flop
DSS	digital signature scheme
ECB	electronic codebook
FG	functional generator
FPGA	field programmable gate array
FT	final transformation
IV	initialization vector
LC	linear characteristic
LCA	linear cryptanalysis

LUA	loop unrolling architecture
MDA	microprocessor design architecture
MDC	manipulation detection code
MAC	message authentication code
NL	nonlinearity
OFB	output-feedback
PA	pipelined architecture
PN	permutation network
RAM	random access memory
RCO	reversible controlled operations
RDDO	reversible data-dependent (data-driven) operations
SCO	switchable controlled operation
SDDO	switchable data-dependent operation (data-driven operation)
SDDP	switchable data-driven permutation
SPN	switchable permutation network
TT	truth table
VBP	variable bit permutation
VLSI	very large scale integration

Notations Used in the Book

- $[x]$ denotes the integer part of x
- $\{0,1\}^g$ denotes the set of all binary vectors $U = (u_1, u_2, \ldots, u_g)$, where $\forall i \in \{1, \ldots, g\}$ $u_i \in \{0,1\}$
- (X, Y) denotes concatenation of the vectors X and Y
- $X \oplus Y$ denotes the bitwise XOR (EXCLUSIVE-OR) operation of the two vectors X and Y: $X, Y \in \{0,1\}^s$
- $X >>> k$ or $X^{>>>k}$ (or $X^{>k>}$) denotes to-right rotation of the word X by k bits
- $X <<< k$ or $X^{<<<k}$ (or $X^{<k<}$) denotes to-left cyclic rotation of the word $X = (x_1, \ldots, x_{32})$ by k bits, i.e., $\forall i \in \{1, \ldots, 32 - k\}$ we have $y_i = x_{i+k}$ and $\forall i \in \{33 - k, \ldots, 32\}$ we have $y_i = x_{i+k-32}$.
- $\varphi(U)$ denotes Hamming weight of binary vector U; $\varphi(U)$ is equal to the number of nonzero components of U, i.e., $\varphi(U) \overset{def}{=} \sum_{i=1}^{g} u_i$
- $X \otimes Y$ denotes bitwise AND operation of the two vectors X and Y: $X, Y \in \{0,1\}^{(g)}$. For $c \in \{0,1\}$ and $X \in \{0,1\}^g$, we define $Y = c{\cdot}X$ where $y_i = c{\cdot}x_i$ $\forall$ $i \in \{1, \ldots, g\}$
- $\varnothing$ denotes a reserved two-place operation
- $\varphi'(U)$ denotes the parity of $\varphi(U)$, i.e., $\varphi'(U) \overset{def}{=} \varphi(U) \bmod 2$
- "$\leftarrow$" or ":=" denotes assign operation
- $\bullet$ denotes the binary scalar product $c = A \bullet X = \varphi'(A \otimes X)$, $c \in \{0,1\}$
- $\mathbf{F}_1||\mathbf{F}_2$ denotes the cascade of the $\mathbf{F}_1$ and $\mathbf{F}_2$ transformation boxes, i.e., $\mathbf{F}_1||\mathbf{F}_2(X_1, X_2) = (\mathbf{F}_1(X_1), \mathbf{F}_2(X_2))$
- $\mathbf{F}^\circ\mathbf{S}$ denotes superposition of **F** and **S** transformations
- "$+_z$" denotes modulo 2^z addition of words (for example, the expression $j \leftarrow W \bmod 2^{11}$ is equivalent to the expression $j \leftarrow Z +_{11} 0$)
- "$-_z$" denotes modulo 2^z subtraction
- $\langle Z \rangle$ denotes the average value of the variable Z
- $\#\{z_1, z_2, \ldots\}$ denotes the number of elements in the set $\{z_1, z_2, \ldots\}$
- $a|b$ denotes that number a divides number b

Chapter 1

Short Introduction to Cryptography

This chapter introduces some basic notions of cryptography and describes several simple hand enciphering systems. It considers application items of cryptography to provide security in telecommunication and computer systems. Block ciphers, their operation modes, and topologies are also described. Hash functions and other cryptographic primitives used to check information integrity, as well as information authentication methods, are introduced. Finally, controlled operations are presented as the main primitive of the data-driven ciphers.

1.1 Symmetric Cryptosystems

1.1.1 Basic Notions

Human history is replete with stories about ciphers' being used to send secret messages. Let us start with terminology. The word *cryptography* comes from the Greek for "hidden writing." *Cryptography* deals with designing methods and algorithms for transforming the source text (*plaintext* or *cleartext*) into a pseudorandom sequence of symbols, which is called *ciphertext* (or *cryptogram*). The set of letters used to write plaintexts is called *source alphabet* (or *plaintext alphabet*). The set of letters used to write cryptograms is called *ciphertext alphabet* (or *output alphabet*). Using cryptography, a "bad guy" (an *opponent*) could intercept a ciphertext message with the intent to access its contents (plaintext). Cryptographic transformation (called *ciphering* or *encryption*) should prevent the possibility that such an unauthorized person could perform reversal transformation (called *deciphering* or *decryption*) of the ciphertext into plaintext, but the legitimate recipient would be able to decipher the cryptogram. Thus, the addressee is able to use something that cannot be used by the opponent. This additional secret information (*secret key* or simply *key*) is shared by the message sender and the recipient.

The ciphers (also called *ciphersystems* or *cryptosystems*) should provide *security*, which means that plaintext extraction from the ciphertext is practically impossible. While using a ciphersystem, our secrets depend on its security. Therefore, designers of the encryption systems should estimate the security of their algorithms. Security estimation is performed analyzing different possible attacks against a cipher. The aim of the cryptographic attack is to discover the key or the plaintext (this is called *breaking the encryption algorithm*), a process also known as *cryptanalysis*. Thus, the design and analysis of ciphers are two indivisible parts of cryptography, although some authors distinguish between *cryptography* and *cryptanalysis*, calling *cryptography* the science of designing ciphers and *cryptanalysis* the art of breaking the cryptosystems. The science comprising both parts is called *cryptology*. The reader should note that the terms *cryptography* and *cryptology* are often used as words having the same meaning, namely, the art and science of designing both encryption methods and techniques for breaking the cryptosystems.

An early user of cryptographic techniques was the famous Roman commander-in-chief and statesman Gaius Julius Caesar. He used extremely simple cipher (called *Caesar's cipher*). Caesar's cipher is composed of the ciphertext alphabet produced by rotation of the source alphabet to the left by three positions (see Table 1.1). Encryption with Caesar's cipher consists in replacing each letter in the plaintext by the corresponding letter of the ciphertext alphabet. This process represents a set of consecutive substitution operations. What is the secret key in Caesar's cipher? We could say that it is the substitution table used (Table 1.1). Suppose that the opponent knows that the ciphertext alphabet is produced using rotation of the plaintext. Then the key would be the number of positions by which the rotation is performed. As we have only 25 different keys in Caesar's cipher, it is too small to provide security, even for beginners in cryptography. If we suppose the opponent does not know what algorithm has been used to create the cryptogram, then we could say that the key is sufficiently large.

Generally, one can use some random permutation of the source alphabet to produce the ciphertext alphabet or denote each letter of the plaintext with a new token (*symbol*), for example, pairs of source alphabet letters or figures (see Table 1.2). Such a cipher is called *monoalphabetical*. This term denotes ciphers that use one alphabet to perform letter substitution while enciphering. In monoalphabetical ciphers we have 26! different keys (this is the number of different permutations that are possible over a 26-letter alphabet). With only a cryptogram produced by the monoalphabetical method, the opponent would be able to easily figure out the ciphertext alphabet. Therefore, in monoalphabetical ciphers that use arbitrary

Table 1.1 Caesar's cipher: a possible substitution

Source alphabet	a	b	c	d	e	f	g	h	i	j	k	l	m	n	o	p	q	r	s	t	u	v	w	x	y	z
Ciphertext alphabet	d	e	f	g	h	i	j	k	l	m	n	o	p	q	r	s	t	u	v	w	x	y	z	a	b	c

Table 1.2 Different variants of the substitution alphabet

Source alphabet	a	b	c	d	e	f	g	h	i	j	k	l	m	n	o	p	q	r	s	t	u	v	w	x	y	z
Ciphertext alphabet 1	c	m	r	z	l	v	q	d	k	e	n	x	s	y	b	j	g	o	f	h	t	a	p	w	u	i
Ciphertext alphabet 2	q	g	p	t	a	d	z	o	e	b	j	h	r	u	y	i	m	v	x	c	k	s	w	f	l	n
Ciphertext alphabet 3	53	87	64	69	51	40	28	39	41	25	93	73	29	43	84	27	72	05	18	29	50	91	37	72	74	93

symbols to define the ciphertext alphabet, we also have 26! different keys. It is a sufficiently large number to prevent a key-exhaustive search attack. However, there are other possible types of attacks that reduce drastically the difficulty of breaking the monoalphabetical cryptosystems. Such attacks are based on the statistics of the language, i.e., the experimental frequency of letters in the plaintext (see Table 1.3). Because a monoalphabetical substitution does not mask the frequency distribution, the symbols occurring with high frequency in the ciphertext correspond to the source alphabet letters having higher frequency in the plaintext. This attack is called *frequency analysis*. The frequency analysis is very efficient in breaking monoalphabetical ciphers.

To make the cryptosystem more secure against frequency analysis, early cryptographers proposed the following types of ciphers:

1. Polyalphabetical cryptosystems
2. Homophonic cryptosystems
3. Transposition cryptosystems

Encryption with the last type system consists of permutation of the plaintext letters. This method retains the frequencies of individual letters; therefore, the simple frequency analysis becomes inefficient.

Polyalphabetical cryptosystems use many different substitution alphabets to perform enciphering plaintexts. They differ from each other in the substitution alphabets used and the rule defining selection of the current substitution alphabet to replace the current source text letter. The source frequency is masked due to the use of different symbols to replace the same source text letter at different encryption steps.

Table 1.3 Relative frequency of letters of the English language

Letter	*Relative frequency f* (%)	*Letter*	*Relative frequency f* (%)
a	8.167	n	6.749
b	1.492	o	7.507
c	2.782	p	1.929
d	4.253	q	0.095
e	12.702	r	5.987
f	2.228	s	6.327
g	2.015	t	9.056
h	6.094	u	2.758
i	6.966	v	0.978
j	0.153	w	2.360
k	0.772	x	0.150
l	4.025	y	1.974
m	2.406	z	0.074

In homophonic cryptosystems, each source-alphabet letter is put into correspondence with a unique subset of the ciphertext-alphabet symbols, each of which is included in only one of the subsets. The subsets are composed so that the cardinal number of a subset is proportional to the frequency of the letter to which the subset corresponds.

Let us use the following notation:

L_i is the source-alphabet letter ($i = 0$ to 25).
f_i is the frequency corresponding to the letter L_i ($i = 0$ to 25).
$\{S_{ij}\}$ is the subset of the ciphertext-alphabet symbols corresponding to L_i ($j = 1$ to N_j, where $N_j = \#\{S_{ij}\}$, the cardinal number of the subset $\{S_{ij}\}$).

While encrypting a plaintext, the current L_i is replaced by an S_{ij} symbol selected at random from the subset $\{S_{ij}\}$. For a homophonic cipher, we have $\#\{S_{ij}\} = const \cdot f_i$. While enciphering $\Pr(S_{ij})$, the probability that the S_{ij} symbol will be used at the current encryption step is equal to

$$\Pr(S_{ij}) = \Pr(L_i) \cdot \Pr(S_{ij}/L_i) = \Pr(L_i) \cdot N_j^{-1} \approx f_i \cdot N_j^{-1} = f_i \cdot (const \cdot f_i)^{-1} = const^{-1} \quad (1.1)$$

Each of the S_{ij} symbols is used approximately with the same probability. Therefore, the source-message statistics are hidden in the cryptogram. The homophonic cryptosystems are significantly more secure than the monoalphabetic ones, but a demerit is the need of a very large output alphabet in a manual enciphering system. In computers or modern telecommunication systems, the demerit of monoalphabetic cryptosystems is the increased size of the ciphertext compared to the plaintext. (Considering their security with regard to standard attacks, we must indicate that they are not secure against simple known plaintext attacks.)

1.1.2 Additive Ciphers

Using Caesar's cipher (see Table 1.1), we can encipher a cleartext letter by replacing it with the letter beneath it. For instance, "text" becomes "***whaw***." Deciphering is equally easy; we must replace a ciphertext letter with the cleartext letter above it. So, for example, "***flskhu***" is deciphered as "cipher." We can assign to each letter its consecutive number and consider numbers 0, 1, … , 25 as letters. Then, Caesar's cipher is described by a very simple formula:

$$L_i = l_{i+3 \bmod 26} = l_i + 3 \bmod 26 \tag{1.2}$$

where L_i and l_i are the *i*th letters in the ciphertext and plaintext alphabets, respectively. Thus, enciphering consists in adding number 3 modulo 26 to each letter of the plaintext. Therefore, we can call the Caesar's cipher an additive cipher with the key $k = 3$. There are 26 different *additive ciphers* possible with the keys $k_i = 0, 1, 2, \ldots, 25$. The key $k_i = 0$ corresponds to the so-called *trivial* cipher, which performs no encryption.

Table 1.4 shows all 26 additive ciphers. We can use them alternately, selecting different substitution alphabets to replace the current letter of the plaintext. For this purpose we will use a keyword (or simply *key*), which should be written below the plaintext as many times as required to assign a key to each letter of the plaintext. While performing plaintext encryption, we select the alphabet that begins with the letter assigned to the current encrypted letter as the current additive cipher, with the help of which we encrypt the current plaintext letter. By means of this simplest class of ciphers, we can clarify two important notions: the *cipher algorithm* and the *key*. We should distinguish between them. The *algorithm* for an additive cipher can be described by Table 1.4. On the other hand, the *key* is the information indicating to us precisely *how the algorithm proceeds* in a particular message. The sender and recipient must first agree on a cipher algorithm and then, before any transmission, choose and exchange their key. Such an additive cipher can be described as follows:

$$L_j = l_j + K_j \bmod 26 \tag{1.3}$$

where L_j and l_j are the *j*th letters in the ciphertext and plaintext, respectively, and K_j is the *j*th letter in the *keystream* that is periodical in the considered case (simple

Table 1.4 Ciphertext alphabets: 26 additive ciphers

a	b	c	d	e	f	g	h	i	j	k	l	m	n	o	p	q	r	s	t	u	v	w	x	y	z
b	c	d	e	f	g	h	i	j	k	l	m	n	o	p	q	r	s	t	u	v	w	x	y	z	a
c	d	e	f	g	h	i	j	k	l	m	n	o	p	q	r	s	t	u	v	w	x	y	z	a	b
d	e	f	g	h	i	j	k	l	m	n	o	p	q	r	s	t	u	v	w	x	y	z	a	b	c
e	f	g	h	i	j	k	l	m	n	o	p	q	r	s	t	u	v	w	x	y	z	a	b	c	d
f	g	h	i	j	k	l	m	n	o	p	q	r	s	t	u	v	w	x	y	z	a	b	c	d	e
g	h	i	j	k	l	m	n	o	p	q	r	s	t	u	v	w	x	y	z	a	b	c	d	e	f
h	i	j	k	l	m	n	o	p	q	r	s	t	u	v	w	x	y	z	a	b	c	d	e	f	g
i	j	k	l	m	n	o	p	q	r	s	t	u	v	w	x	y	z	a	b	c	d	e	f	g	h
j	k	l	m	n	o	p	q	r	s	t	u	v	w	x	y	z	a	b	c	d	e	f	g	h	i
k	l	m	n	o	p	q	r	s	t	u	v	w	x	y	z	a	b	c	d	e	f	g	h	i	j
l	m	n	o	p	q	r	s	t	u	v	w	x	y	z	a	b	c	d	e	f	g	h	i	j	k
m	n	o	p	q	r	s	t	u	v	w	x	y	z	a	b	c	d	e	f	g	h	i	j	k	l
n	o	p	q	r	s	t	u	v	w	x	y	z	a	b	c	d	e	f	g	h	i	j	k	l	m
o	p	q	r	s	t	u	v	w	x	y	z	a	b	c	d	e	f	g	h	i	j	k	l	m	n
p	q	r	s	t	u	v	w	x	y	z	a	b	c	d	e	f	g	h	i	j	k	l	m	n	o
q	r	s	t	u	v	w	x	y	z	a	b	c	d	e	f	g	h	i	j	k	l	m	n	o	p
r	s	t	u	v	w	x	y	z	a	b	c	d	e	f	g	h	i	j	k	l	m	n	o	p	q
s	t	u	v	w	x	y	z	a	b	c	d	e	f	g	h	i	j	k	l	m	n	o	p	q	r
t	u	v	w	x	y	z	a	b	c	d	e	f	g	h	i	j	k	l	m	n	o	p	q	r	s
u	v	w	x	y	z	a	b	c	d	e	f	g	h	i	j	k	l	m	n	o	p	q	r	s	t
v	w	x	y	z	a	b	c	d	e	f	g	h	i	j	k	l	m	n	o	p	q	r	s	t	u
w	x	y	z	a	b	c	d	e	f	g	h	i	j	k	l	m	n	o	p	q	r	s	t	u	v
x	y	z	a	b	c	d	e	f	g	h	i	j	k	l	m	n	o	p	q	r	s	t	u	v	w
y	z	a	b	c	d	e	f	g	h	i	j	k	l	m	n	o	p	q	r	s	t	u	v	w	x
z	a	b	c	d	e	f	g	h	i	j	k	l	m	n	o	p	q	r	s	t	u	v	w	x	y

repetition of the keyword). The ciphers using a keystream are called *stream ciphers*. Different stream ciphers differ in the method of the keystream generation. In secure stream ciphers, the keystream is indistinguishable from a random sequence of tokens K_j selected from a defined alphabet. If the keystream generation method defines dependence of the current key value K_j on one or more previous letters, L_z or l_z (or both), where $z < j$, we have a stream cipher with *autokey*.

The algorithm and the key have essentially different functions. Usually, the algorithm is rather large. Many of them are based on complex mathematical procedures and are realized by an electronic device. Therefore, the algorithm cannot be kept secret. This means that the overall security of a cryptosystem relies on the secrecy of the key. (Remember the following *Kerckhoffs' principle*: The security of an encryption system should depend on keeping secret only the key, not the transformation algorithm. In other words, the security of a cryptosystem should be evaluated on the assumption that the opponent knows all details of the algorithm, and only the key is a secret element.)

We suppose that a "bad guy" who wants to read our message can, with relative ease, obtain the algorithm, and we are able to keep the key secret; i.e., we suppose bad guys do not know the current key. Thus, the key is to be transmitted securely, and we need to use secure channels. This is the *first fundamental requirement* of cryptography, with which the *secret key distribution problem* is connected.

Why not simply transmit the message using this secure channel? The following arguments favor using the cryptographic transformation:

- Usually, messages are very long (sometimes we need to encrypt the total information on a hard disk or on a flash memory). In contrast, the key is usually small in length (it is as short as justifiable with respect to security). Therefore, the effort in securely transmitting the key is very minimal and the probability that the bad guy intercepts the key relatively small.
- The sender and recipient may choose the moment of key exchange. The exchange can take place days, even weeks, or months before the transmission of the confidential message. In contrast, the message often has to be transmitted at a time that is beyond the sender's control. (Think of political events, unexpected developments in the stock market, and so on.)

The second fundamental requirement of cryptography is the *key authentication*. The receiver should be convinced of the authenticity of the person with whom the secret key is shared. In symmetric-key cryptography, key authentication is performed, and the key distribution problem solved, simultaneously. Therefore, sometimes we do not take into account the second fundamental problem, i.e., the key *authentication problem*. This becomes of prime importance in so-called public-key cryptosystems, which provide excellent solutions to the secret key distribution problem. Using the public-key cryptosystems, one can exchange a key even without a secret channel.

Example of ciphering with Table 1.4

t	h	i	s	i	s	a	p	l	a	i	n	t	e	x	t
s	e	c	r	e	t	s	e	c	r	e	t	s	e	c	r
l	***l***	***k***	***j***	***m***	***l***	***s***	***t***	***n***	***r***	***m***	***g***	***l***	***i***	***z***	***k***

This example shows how the source text "this is a plaintext" is transformed into the ciphertext "***llkjmlstnrmglizk***," with the keyword "secret" used.

1.1.3 Application in Telecommunications and Computer Systems

The importance of cryptography goes far beyond providing data secrecy. As data transmission and processing become more intense, cryptographic methods gain greater importance. Moreover, new information technologies are based on public-key cryptography, which is a comparatively new branch of cryptography. Public-key cryptosystems provide the possibility of implementing protocols in which the interacting parties do not trust each other, because the secret key in such cryptosystems is only known to a single user. Here are some modern applications of cryptography:

- Protection against unauthorized access to information
- Information integrity control
- User authentication
- Information authentication (digital signatures)
- Computerized secret voting
- Digital cash

The first application has been discussed previously. Usually, a valid receiver can easily detect that a cryptogram has been modified or substituted, analyzing the semantics of the decrypted message. However, it is generally desirable to have a formal procedure of data integrity control. For example, when digital data is modified, it is difficult to check the information integrity by means of just semantics. Simple and secure data integrity control methods are based on different kinds of checksums that are transmitted with the clear or encrypted message. A checksum is some additional information, depending on the message, generated by a cryptographic algorithm. The following two types of cryptographic checksums are used:

- Checksums calculated based on a secret key (keyed checksums)
- Checksums calculated without using secret elements (unkeyed checksums)

Cryptographic checksums are often called hash functions. An unkeyed hash function is also called *manipulation detection code* (MDC). MDC algorithms are often constructed based on block ciphers that are used to consecutively process fixed-size data blocks into which the message is divided. The secure size of MDC is 128 bits and more. Secure keyed checksums have smaller length than secure unkeyed checksums. A keyed hash function is called *message authentication code* (MAC). The MAC is computed using a cryptographic algorithm that specifies how the MAC value depends on each bit of both the message and the secret key. The longer the MAC, the higher is the probability that message manipulation will be detected by the authorized receiver. An opponent can modify the message, but he or she cannot calculate a new MAC value corresponding to the modified message, because the latter requires use of the secret key. The opponent either doesn't change the MAC or attaches a new MAC value; the probability that the manipulation won't be detected is $P = 2^{-m}$, where m is the MAC length in bits. The secure size of MAC is 64 bits and more.

Unkeyed checksums are required in many applications such as *digital signature schemes* (DSSs). In some other cases, it is also necessary to use cryptographic-checksum algorithms, which are more sensitive to minor changes in messages, to provide the best protection with an unkeyed algorithm. A special unkeyed checksum-like function is called a *one-way hash function* (or simply *hash function*). A hash function should be designed so that it is computationally impossible to construct a forged message that yields the same result as a true message. Hash functions are usually used as part of the digital signature systems. A hash function is a computationally efficient function mapping binary strings of arbitrary length to binary strings of some fixed length, called *hash values*: $\mathbf{Hash}(M, H_0) = H$, where $M = (M_1, M_2, \dots, M_n)$ is a hashed message, H_0 a specified initial value, and H is the hash value from M. Because H_0 is a specified value, one can simply write $\mathbf{Hash}(M) = H$. Usually, the hash function has an iterated structure. An iterated hash function, **Hash**, is determined by an easily computable round function, **h**, from two bit strings M_i and H_{i-1} of respective lengths m and h:

$$H_i = \mathbf{h}(M_i, H_{i-1}) \tag{1.4}$$

where $i = 1, 2, \dots, n$ and $H = H_n$. Hash function should be designed so that it satisfies the following two requirements:

1. It is computationally impossible to construct a forged message that yields the same result as a true message (given a message M and its hash value H, it should be computationally difficult to find a message $M' \neq M$ that hashes to the same value H).
2. It is computationally impossible to construct two different messages that yield the same result (it should be computationally difficult to find two different messages M and $M' \neq M$ that hash to the same value, i.e., $H' = H$).

The first requirement means that the hash function should be a "one-way" function. The second means that the function should also be "collision free." A number of hash functions have been proposed earlier. Some well-known hash functions are Snefru, N-Hash, MD5, MD4, MD2, and SHA [30,52,60]. Different hash functions can be implemented using 128-bit block ciphers as the main unit of the round hash function.

Computer systems, which cannot differentiate between requests for service by legitimate users and unauthorized access attempts, are vulnerable to a variety of attacks. One of the important security mechanisms is authentication, which is aimed at establishing the validity of a message, or verifying the eligibility of a user to receive specified privileges when entering a computer system. When entering the system, the user is to be *identified*, generally by means of a unique user name or number. The identification process enables the computer system to recognize an entity. Identification is used only to indicate what current entity is to be *authenticated*. Authentication mechanisms are critical to the security of any automated system. If the identity of legitimate users is verified with a certain degree of accuracy, then those attempting to gain access without proper authorization can be denied permission to use the system. After verifying the legitimate user's identity, specified privileges can be allowed to that individual. Without verifying the identity of users, a computer system will not be able to protect itself against unauthorized access. Access control techniques incorporate some kind of authentication mechanism.

A number of different methods have been proposed, and are available, for performing user authentication. Authentication methods can be divided into three main categories. The first category includes methods based on something the user knows (for example, a secret key or a password). The second category includes methods based on something the user possesses (for example, an authentication token). The third category (biometric authentication of human users) includes methods based on physical characteristics of the individual (for example, a fingerprint or voice pattern). User authentication based on the fact that legitimate users possess some information not available to outsiders is preferable in many practical applications, for example, in computer security systems for mass assignment. For special applications, one can use a combination of these variants. The third variant demands some expensive equipment to be used. For verification of the identity of users, computer security systems use software, hardware, or a combination of both. The first variant is well suited for implementation with software. The strength of an authentication mechanism of the first type depends on how much of the secret information available with the user is used to perform authentication. The larger the key or passwords used, the more secure is the authentication mechanism. The required strength of an authentication mechanism should be decided based on the security requirements of the application, the environment in which the system is to be utilized, and the system services to be provided.

An authentication mechanism based on passwords differs from one based on secret keys. The first is used to control logging into a workstation. Passwords are not explicitly stored in the computer memory. This prevents a possible inside adversary from reading legitimate users' passwords. For a security system to be able to identify legitimate users, some password images are stored in the computer memory; these are computed according to a special cryptographic algorithm that implements a so-called *one-way function*, $y = F(x)$, which can be easily computed provided that the argument x is given. The main requirement of this function is that the complexity of computing the argument at which the function takes on a given random value is extremely high.

The user authentication protocol using a password is as follows:

- The system requests identifier.
- The user enters his or her username UN.
- The security system requests for the password.
- The user enters his or her password Password_UN′.
- The system computes the value $y' = F(\text{Password\_UN}')$. Then it compares the $F(\text{Password\_UN}')$ value with the password image value $y = F(\text{Password\_UN})$ stored in the system storage.

If $y' = y$, then the system provides the user access rights corresponding to the UN identified. This example describes the user authentication protocol on a workstation. For mutual authentication of remote workstations, protocols based on secret keys and using random values are employed. In the following example, the **E** enciphering algorithm and the K secret key are shared by remote stations A and B:

- The A workstation sends an initiating message to B workstation.
- The B workstation sends a random number R to A workstation.
- The A workstation encrypts the R value using the secret key K: $C_a = \mathbf{E}_K(R)$. Then it generates a random value R' and sends the values R' and C_a to B workstation.
- The B workstation computes the $C_b = \mathbf{E}_K(R)$ value and compares C_b with C_a. If $C_b = C_a$, it concludes that it receives the messages from A. Then it computes the $C_b' = \mathbf{E}_K(R')$ value and sends the C_b' value to A.
- The A workstation computes the $C_a' = \mathbf{E}_K(R')$ value and compares C_a' with C_b'. If $C_a' = C_b'$, it concludes that it receives the messages from B.

This scheme of mutual authentication of two remote parties is called a *handshake protocol.*

Information authentication is a process that establishes the fact that a received message was sent by an authorized sender or even signed by him or her. In symmetric cryptosystems, authentication is performed using MAC checksums calculated by means of a shared secret key. DSSs based on symmetric encryption require the third party to participate in the DSS protocols. DSSs based on public-key cryptosystems are more efficient. They do not require any trust party to participate in the signature generation and signature verification processes.

DSS protocols are based on public-key algorithms that involve the use of two keys—a *public* and a *private* one. With the private key, it is possible to generate a message with a special internal structure related to the document being signed and to the public key. With a public key (which is known to all users of a cryptosystem and to a potential attacker), everybody is able to check the validity of the digital signature. A valid signature can be formed only by means of the private key, which is known solely to one entity, i.e., a user to whom the corresponding public key is assigned. The use of a private key to generate a digital signature for an electronic document is verified with a *public key*. The public key is useless for valid signature generation, although it is derived from the private key.

1.1.4 Block Ciphers

In information security practice, data confidentiality (privacy or secrecy) is supported by the use of symmetric-key encryption systems. There are two known classes of such systems: (1) stream and (2) block ciphers. Stream ciphers encrypt individual bits or characters of an input plaintext (or ciphertext), one at a time, using a simple encryption transformation. Usually, the transformation consists in XORing one bit or word (having a length of 8, 16, or 32 bits) of plaintext with the respective bit or word of the keystream. The type of a stream cipher is defined by the algorithm generating the keystream. With the stream cipher, the same plaintext words are assigned to different ciphertext words, every time they are encrypted. Usually, stream ciphers are faster than block ciphers and have limited or no error propagation; therefore, they are preferable (or even mandatory) in some telecommunication applications. However, the most widely used secret key cryptosystems are block ciphers; these ciphers are often preferable for data encryption in computer and telecommunication systems. Block ciphers are cryptosystems that encrypt information in blocks of a fixed length, for example, n bits. This type of cryptographic transformation is called *block ciphering* or *block encryption*. In block ciphering, a message is represented as a series of n-bit blocks. Usually, the messages have an arbitrary length, generally not a multiple of the block length. The following method of complementing the last data block can be used.

The last data block is often complemented with a binary vector $(1, 0, 0, \dots, 0)$, in which the number of zeros can be anywhere from 0 to $(n - 2)$. If the length of the last data block is equal to n bits, an additional n-bit block $(1, 0, 0, \dots, 0)$ is appended to the message. This method makes it possible to determine uniquely the appended binary vector and cut it if necessary. Using this method, one can represent any message, M, as a concatenation of n-bit subblocks, M_i:

$$M = M_1||M_2|| \dots ||M_i|| \dots ||M_m \tag{1.5}$$

Each data block can be transformed independently of the other blocks; therefore, while using block ciphers, direct access to the encrypted data blocks is possible.

Usually, the block-enciphering algorithms transform any given plaintext block into arbitrarily given n-bit ciphertext block, depending on the key used. Generally, the output block length is equal to or greater than the input block length. A ciphertext block cannot be of less size than the plaintext block, because in that case several different plaintext blocks would correspond to the same ciphertext block (which would result in ambiguous decryption). If the length of the output block is greater than n, then several different ciphertext blocks will correspond to the same plaintext block. In such cases, unique decryption is possible. (Examples of such cryptosystems are homophonic ciphers.) However, in automated systems, it is not desirable for the encrypted message to be of a larger size than the plaintext message. Therefore, in most commonly used block ciphers, the size of output blocks is equal to that of input blocks. Block encryption algorithms specify a one-to-one correspondence between the set of possible input blocks and that of possible output blocks. As input and output block sets coincide, encryption defines a substitution on the $0, 1, \ldots, 2^n - 1$ set of numbers, which can be presented as follows:

$$\begin{pmatrix} 0 & 1 & 2 & \ldots & 2^n - 1 \\ \mathbf{E}_K(0) & \mathbf{E}_K(1) & \mathbf{E}_K(2) & \ldots & \mathbf{E}_K(2^n - 1) \end{pmatrix} \tag{1.6}$$

where $\mathbf{E}_K(M)$ is an encryption function controlled with the key K. The key specifies what kind of substitution table is implemented by the encryption algorithm. Actually, it is impossible to construct such a table because of the large value of n (usually for practically used ciphers we have n = 64 or 128 bits). However, selecting an algorithm and a key, we define a substitution table that sets the correspondence between plaintext blocks, M, and ciphertext blocks, $C = E_K(M)$. For a given key, one substitution is implemented. In general, different substitution tables correspond to different keys. If the length of secret key $|K| = k$ bits, then the block cipher specifies no more than 2^k different substitutions. This is only a very small portion of all possible substitutions numbering $2^n!$; to implement all of these with the same algorithm, one needs to use a key of the length

$$k = \log_2(2^n!) \approx n2^n \text{ bits} \tag{1.7}$$

Thus, the block encryption is a monoalphabetic substitution over a very large alphabet. Because of the large length of the alphabet, the frequency cryptanalysis is not efficient against block ciphers. In the case n = 32, the frequency cryptanalysis becomes extremely complex, although block ciphers with a 32-bit input data block can be used sometimes. The minimum value of n is considered to be 64 bits. The larger the input block size, the higher is the security that can be potentially obtained. However, the larger the input block size, the greater the hardware implementation cost. Therefore, at present 64- and 128-bit block ciphers are usually used, although sometimes 4096-bit algorithms are also used (for example, to perform hard-disk encryption). Input data blocks of sufficiently large size are only the first security requirement of block ciphers. The second is a secret key of sufficiently large size. At present, a value of $k \geq 128$ bits is recommended. The third requirement is the use of secure ciphering algorithms. To evaluate security of a block algorithm, all known cryptographic attacks should be considered. Security is the property of a cipher's resistance to different attacks. The objective of an attack is to recover plaintext from ciphertext or to deduce the key. Usually, the following main types of attacks are considered, in which the transformation algorithm is presumably known:

- A ciphertext-only attack: This is one in which the cryptanalyst tries to recover the key or the plaintext, analyzing only the ciphertext.
- A known-plaintext attack: Here, the cryptanalyst knows the quantity of plaintext and its correspondence to a quantity of ciphertext. The person tries to recover the key or the plaintext corresponding to some other portion of the cryptogram.
- A chosen-plaintext attack: The cryptanalyst chooses a quantity of plaintext and is then given the corresponding ciphertext.
- A chosen-ciphertext attack: The cryptanalyst chooses a quantity of ciphertext and is then given the corresponding plaintext.
- An adaptive chosen-plaintext attack: The cryptanalyst chooses a plaintext that is encrypted, then analyzes the ciphertext corresponding to the chosen plaintext, and selects another plaintext based on the results of the previous encryption. The individual can perform an unlimited number of such steps.
- An adaptive chosen-ciphertext attack: In this form of attack, the cryptanalyst chooses a ciphertext that is decrypted, then analyzes the plaintext corresponding to the chosen ciphertext, and selects another ciphertext based on the results of the previous decryption. An unlimited number of such steps can be performed.

Each type of attack can be implemented in different concrete forms, depending on the concrete type of encryption algorithm that is analyzed. At present, a cipher is considered safe if it is secure against adaptive chosen-text attacks, i.e., all listed attacks. Sometimes other types of attacks are considered—for example, the *chosen-key attack*, in which the attacker has some knowledge about the relationship between different keys. The most important types of chosen-text attacks (including adaptive ones) against block ciphers are linear and differential cryptanalysis.

The Data Encryption Standard (DES) algorithm with a 64-bit data block was a leading block cipher for a long time, supporting a lot of security products. It was a sample cipher for many block cipher designers and the most known target for cryptanalysts. The results of exploiting and testing DES are brilliant for this cipher: (1) The best practical attack is an exhaustive search for the secret key; (2) it provides comparatively high performance with cheap hardware, or software; (3) users trust DES as secure and free of trapdoors cipher. In the mid-1990s, the power of personal computers had reached such a level that further use of DES became dangerous because of its very small key (56 bits). A number of new 64-bit block ciphers with larger key size had been proposed: FEAL (64 bits), GOST (256 bits), IDEA (128 bits), LOKI91 (64 bits), RC5 (variable key length), MISTY (128 bits), different variants of SAFER (64 or 128 bits), Blowfish (≤448 bits), and some others. Some of them were introduced in practical use on a broad scale and gained recognition as secure ciphers, whereas others were criticized. At the same time, the triple DES (112 or 168 bits) had been introduced as an alternative 64-bit cipher based on trusted design.

The Blowfish is interesting in its use of very large key-dependent S boxes. The RC5 is a very interesting parameterized solution in which the length of the key and data block and number of rounds are variable. A design with a 128-bit data block was proposed in SQUARE. To perform fast disk sector encryption, COBRA ciphers (64- or 4096-bit block size) based on data-dependent subkey selection had been proposed. Except for a few specific designs, most of the proposed ciphers can be attributed to several main architectures arising from Shannon's concept of product ciphers, corresponding to which different kinds of simple and insecure ciphers form a superposition, representing secure cryptosystems called product ciphers [67]. Shannon argued that simple transformations are efficient to compose a secure product cipher, if they combine confusion and diffusion, mixing the key with data.

The substitution–permutation networks (SPNs) are an example of the product cipher. Some SPNs represent a set of consecutive alternating layers of P and S boxes. The P boxes have a large size (equal to the data block size) and fixed topology of input–output connections. P boxes are implemented in hardware as simple wiring, whereas their software implementation is expensive, if arbitrary permutation is predefined. The layers of the S type contain many S boxes of comparatively small size. They are efficient in software implementation; in hardware they usually determine the implementation cost. The larger the S boxes, the more expensive the implementation. S boxes with 8-bit input (8 × 8 S boxes) are considered large. The S layers of SP-based ciphers define the nonlinear part of the transformation, having confusion properties. The P layers represent the linear part of the transformation and contribute diffusion. If the P and S layers are selected properly, then SPNs have the following properties:

- Avalanche property: A single input bit change forces the complementation of each output bit with probability approximately equal to 0.5.
- Completeness property: Each output bit is a complex function of every input bit.

The key can be mixed with data using some simple linear operations between the P and S layers. The key addition layer is used conventionally in most of block ciphers. Another solution is to use key-dependent P or S boxes; however, the key-dependent operations should be designed very carefully (to define security against related key attacks).

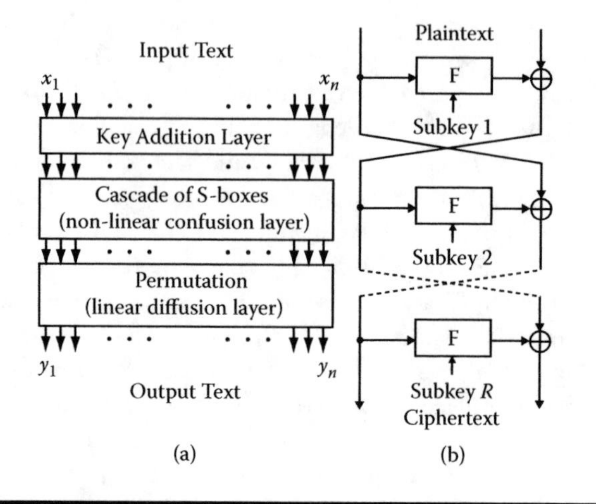

Figure 1.1 One round of uniform SPN (*a*) and *R*-round Feistel's network (*b*).

In general, encryption and decryption algorithms of the product cipher are different. This increases hardware implementation costs. Feistel proposed an elegant topology of SPNs, which defines the same algorithm for both encryption and decryption. In Feistel's network, the data block is divided into two subblocks that are transformed alternately, not in parallel. This feature imposes some restrictions on the performance of such ciphers. The consecutive transformation of two data subblocks decreases the amount of parallelism but increases the propagation of local changes. In SPNs having the usual topology, the whole data block (it can be divided in subblocks though) is transformed simultaneously. Most of the well-known ciphers can be divided into two groups:

1. Ciphers based on uniform SPNs (for example, SAFER, SHARK, 3-WAY, and SQUARE; the general structure of the encryption scheme is presented in Figure 1.1*a*).
2. Ciphers based on Feistel's network (for example, DES, GOST, Blowfish, CAST, FEAL, LOKI91, Knafre, MISTY1, and MISTY2; the general structure of the Feistel-like cryptosystems is presented in Figure 1.1*b*, in which the round function F defines arbitrary mapping and can be implemented as, for example, in some SPNs).

One of the features of Feistel's network is that the round function F need not be invertible. For the arbitrary F function, the round transformation is an involution. The Feistel's cipher approach can be executed in different ways, for example:

- By dividing the input into several parts having the same size
- By dividing the input into two or more parts of different size

Because of DES there is much cryptanalytic experience available with Feistel's ciphers. In general, no certain conclusion can be drawn on which of the previously mentioned two structures of SPNs is the best.

In the mid-1990s, it became evident that DES was an out-of-date encryption algorithm because of its small key (56 bits), and many new block ciphers were proposed comprising a number of design elements. At the same time, theoretic research resulted in the formulation of main requirements for block ciphers to be secure against several variants of cryptanalysis, in the first place, against differential and linear attacks. In 1997, the National Institute of Standards and Technology (NIST) initiated a process of selecting a new 128-bit block encryption algorithm to replace DES. Leading cryptographers responded to the call and proposed a variety of new designs. In 1998, NIST announced the acceptance of 15 candidates for the new advanced encryption standard (AES) [1]. The selection of AES was a two-year public process, and the cryptographic community participated in it widely. A lot of security research results have been published on each of the 15 candidates. Besides, all of them were studied regarding their efficiency characteristics in hardware and software implementation, and suitability to different applications. NIST reviewed the results and selected five algorithms (called AES finalists) for further consideration: MARS, RC6, Rijndael, Serpent, and Twofish. At the next step of the AES competition, after more detailed investigation of the finalists, the Rijndael algorithm, developed by Joan Daemen and Vincent Rijmen, has been announced as the AES (October 2000). The reasons for Rijndael to be selected as AES are the following:

- It is suitable for secure implementation and provides good security margin. The security margin is defined as follows. If the minimum number of rounds needed for a cipher to be secure is measured as R_{min} and the number of specified rounds is R, then the security margin is measured as $100\% \cdot (R - R_{min})/R_{min}$.
- It was the only AES finalist that supports different block lengths—128, 192, and 256 bits.
- It suits both the hardware and the software implementation well, providing high performance.

Detailed description of the AES finalists and different implementation and security items relating to each are presented in a number of papers and books. As one of the important results of the AES competition, new theoretic ideas have come into practical designs, assisting further progress in the block cipher synthesis. However, the AES competition was oriented to the block cipher design, whereas the practical needs for cryptographic algorithms are wider. As an attempt to cover this gap, a new research project, "New European Schemes for Signature, Integrity, and Encryption" (NESSIE), was initiated by the Information Societies Technology (IST) Programme of the European Commission (for detailed information see

Web site http://cryptonessie.org) in January 2000. The goal of the NESSIE project was to put forward a portfolio of efficient cryptographic primitives obtained after a transparent and open evaluation process concerning security and performance aspects. The main features of the NESSIE project, as compared to the AES competition, are the following:

- The NESSIE had announced many different nominations: block ciphers, synchronous stream ciphers, self-synchronizing stream ciphers, message authentication codes, collision-resistant hash functions, one-way hash functions, families of pseudorandom functions, asymmetric encryption schemes, asymmetric digital signature schemes, and asymmetric identification schemes.
- It was also announced that NESSIE would consider primitives designed for use in specific environments.
- The main goal was to select the best ideas and designs rather than propose a new European standard on some of the nominations.

The NESSIE project initiated further progress in block cipher design; however, in both the AES and the NESSIE projects, data-dependent (DD) operations (DDOs) with a large number of realizable modifications have not been presented in the cipher design, although the DDO appears to be a very attractive new primitive. The main feature of the RC5, RC6, and MARS ciphers is the use of variable rotations with 32 modifications, which are implemented as DD rotations. DD permutations (DDPs) have been proposed as DDOs with a very large number of modifications, with the permutation networks (PNs) being used as DDP boxes. In this book, we will pay significant attention to DDP-based block ciphers. Before a detailed description of concrete ciphers, it is logical that block ciphers' modes of operation are considered.

There are different known methods of employing block ciphers (modes of operation) [52, 60]. In the *electronic codebook* (ECB) mode, a block cipher encrypts independently fixed-size n-bit data blocks. The ECB mode has disadvantages in many applications, motivating other cryptographic modes. An encryption mode combines the basic cipher with some sort of feedback and some simple operations. In general, the security provided by data encryption is a function of the underlying block cipher, and not the mode, for which the following requirements are formulated:

- The encryption mode should not compromise the security of the underlying cipher.
- The mode should not be significantly less efficient than the underlying encryption algorithm.
- The mode should not increase the size of the encrypted message.

A cryptographic mode usually combines the basic block cipher, some sort of feedback, and some simple operations (sophisticated operations are not required, because security is a function of the underlying block cipher).

The ECB mode is the most natural and evident way to use a block cipher: A block of plaintext is encrypted into a block of ciphertext. This is the easiest mode to work with. Each plaintext block is encrypted independently. You can encrypt and decrypt files that are accessed randomly, for example, a database. If a database is encrypted by the ECB mode, then any record can be added, deleted, encrypted, or decrypted independently of any other records. Another advantage of this mode is the possibility of parallelizing the encryption processes. With multiple encryption processors, different blocks can be encrypted or decrypted without concern about other blocks. In hardware implementation, you can use pipeline iterative architecture. However, the problem with the ECB mode is that some of ciphertext blocks can be transposed, added, or removed.

The *cipher block chaining* (CBC) mode uses a mechanism that defines all previously encrypted data blocks' influence on the current ciphertext block. Chaining adds a feedback mechanism to a block cipher: the results of the encryption of previous blocks are fed back into the encryption of the current block. The result is that each output ciphertext block is dependent not just on the input plaintext block, but on all the previous plaintext blocks. In the CBC mode, the plaintext is XORed with the previous ciphertext block before it is encrypted. Figure 1.2 (encryption) shows CBC encryption in action. After a plaintext block P_i is encrypted, the resulting ciphertext C_i is also stored in a feedback register.

Before the next plaintext P_{i+1} block is encrypted, it is XORed with the feedback register to become the next input to the encrypting routine. The resulting ciphertext is again stored in the feedback register, to be XORed with the next plaintext block, and so on, until the end of the message. Thus, the encryption of each block depends on all the previous blocks.

Decryption is just as straightforward (see Figure 1.2). A ciphertext block is decrypted normally and also saved in a feedback register. After the next block is decrypted, it is XORed with the results of the feedback register. Then the next

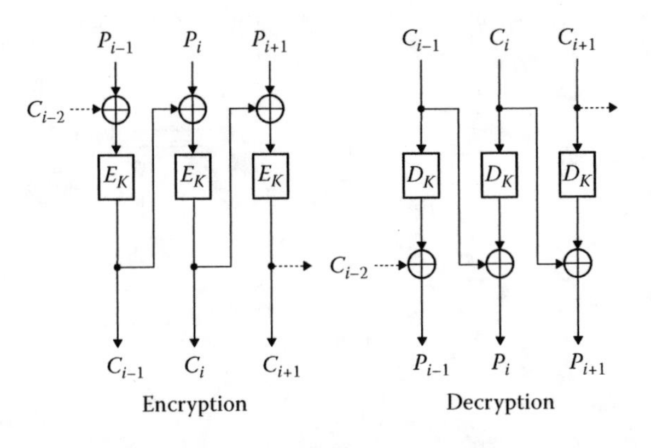

Figure 1.2 Cipher block chaining mode.

ciphertext block is stored in the feedback register, and so on, until the end of the message. Using formulas, this mode is described as follows:

$$C_i = E_K(P_i \oplus C_{i-1}) \tag{1.8}$$

$$P_i = C_{i-1} \oplus D_K(C_i) \tag{1.9}$$

In the CBC mode, identical plaintext blocks are transformed into different ciphertext blocks only when some previous plaintext block is different. Two identical messages will still transform into the same ciphertext. This can be prevented by using an *initialization vector*, that is, a random data block that is used to add to the first data block.

In the *cipher-feedback* (CFB) mode, block ciphers are used to generate a keystream, i.e., a sequence of tokens (symbols, binary vectors, or bits) dependent on a secret key and some initializing vector. Actually, in this mode the block cipher is used as a self-synchronizing stream cipher. With the CFB mode, the block cipher is able to encrypt data by small portions. This is very useful in some network applications. In a secure network environment, for example, a terminal must be able to transmit each character to the host as it is entered. In the CFB mode, data can be encrypted in units smaller than the block size. If the size of these units is t bit, we call the mode t-bit CFB. For example, 8-bit CFB encrypts one ASCII character at a time. In general, t = 1 to n, where n is the block cipher input size. Figure 1.3 shows an 8-bit CFB mode of the encryption process with a 64-bit block cipher. To encrypt the current character of the plaintext, the block algorithm in the CFB mode generates an 8-bit keystream token. The content of the shift register is used as an input block, which is transformed into an output 64-bit block. The leftmost byte is used as the keystream token while the current ciphertext symbol c_i is produced. Then, the content of the shift register is updated; i.e., the register content is shifted to the right by 8 bits, and the c_i value is fed into the rightmost byte of the register. Then

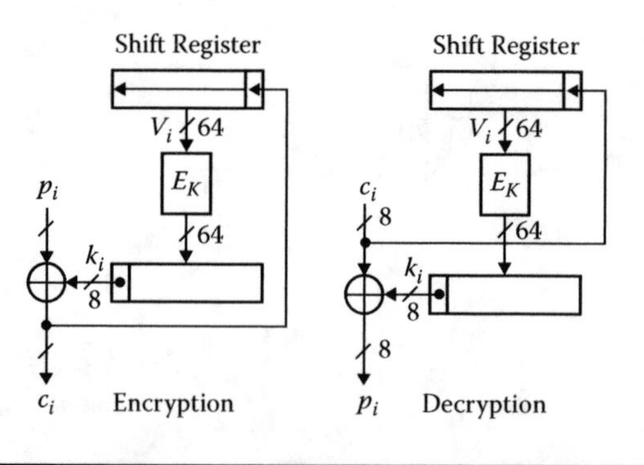

Figure 1.3 Cipher-feedback mode.

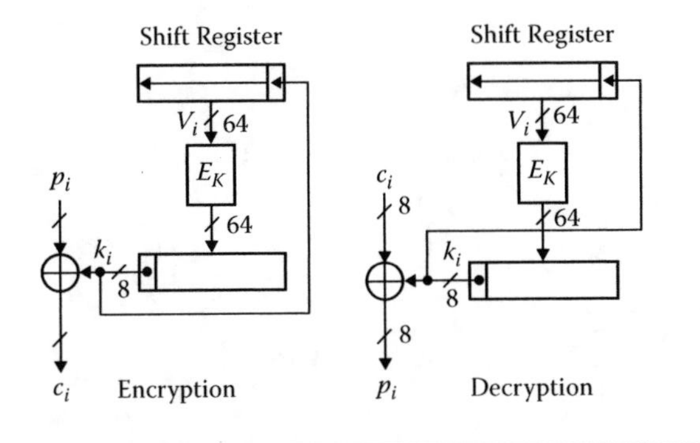

Figure 1.4 Output-feedback mode.

the next plaintext character is encrypted in the same manner. Decryption is the reverse of the encryption process. On both the encryption and the decryption sides, the block algorithm is used in its natural encryption mode. As with the CBC mode, the CFB mode links the plaintext characters together so that the current ciphertext character depends on all the preceding plaintext characters, and on the current one. To initialize encryption with the CFB mode, some initial value should be fed into the shift register. This value can be specified, or it can be chosen arbitrarily and sent to the receiver together with the cryptogram. The randomly generated initial value is called *initialization vector* (IV). The IV value must be unique and need not be secret, as with the IV used in the CBC mode. The IV value can be deterministically changed with every message. For example, one can use a serial number of the message as the IV value, which increases after each message and does not repeat during the lifetime of the key.

Output-feedback (OFB) mode is a method of running a block cipher as a synchronous stream cipher (sometimes this mode is called *internal feedback*). It is similar to the CFB mode, except that the current keystream token is used to update the content of the shift register before generation of the next keystream token (see Figure 1.4). The OFB shift register must also be initially loaded with a unique (but not secret) initialization vector.

In the *counter mode*, a block cipher is used to encrypt consecutive numbers generated by a counter. Instead of using the ciphertext block to update the shift-register content, the input data block of the block cipher is the output of the counter. After each encryption of the shift-register content, the counter increases by some constant value, typically one. The initialization vector can be used to set the initial value of the counter or add with the counter output (see Figure 1.5).

1.1.5 Controlled Operations as a Cryptographic Primitive

The conventional primitives of symmetric cryptosystems are the following: substitutions and permutations, arithmetic and algebraic operations, as well as some

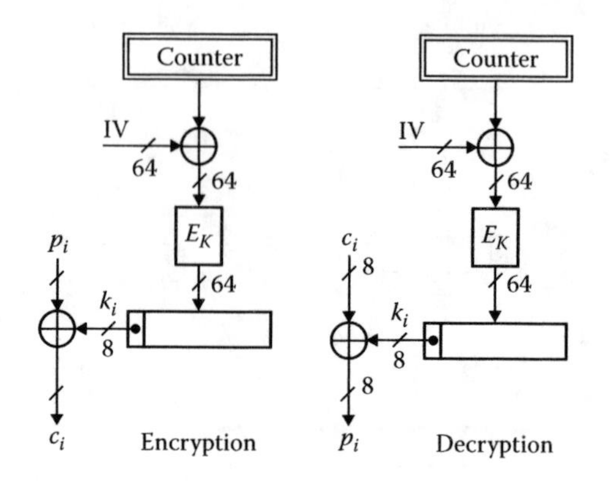

Figure 1.5 Counter mode.

auxiliary operations. The most extensively used primitive is substitution, which is also the most general invertible operation. Security of most block ciphers is based on the use of a certain kind of substitution. The substitution operation is usually implemented as follows:

- In case of software or firmware implementations of the encryption algorithm, the substitution operations are implemented as table lookups. In more detail, the input value is used to address the substitution table and extract from it an output value. Tables defining substitutions are stored in the random access memory (RAM). Therefore, it is easy to implement arbitrary substitution over n-bit binary vectors, where $n = 4$ to 16 bits (such operations are called n-bit substitutions and denoted as $n \times n$ substitutions). For $n = 8$ bits (or $n = 16$ bits), the required memory is 256 bytes (or 128 Kbytes). Implementation of substitutions over vectors larger than 16 bits is problematic, even taking into account the large RAM size of modern computers.
- In case of hardware implementation, substitution operations are implemented with complex electronic circuits. Arbitrary substitutions over binary vectors larger than 8 bits are very hard to implement; i.e., their hardware implementation cost is extremely large.

If our implementation method provides the ability to implement arbitrary-type substitutions, then it is possible to use the best substitutions that comply with certain cryptographic criteria. In case of substitutions of small size (say, 4×4), the most efficient substitutions can be easily found. However, for $n \geq 8$, it is not sufficiently easy to justify selection of the n-bit substitutions for designed cryptoschemes; however, the higher the n value, the more efficient the substitutions. In some designs of block ciphers, the substitutions (they are also called S-box operations) are defined by means of some arithmetic operations possessing certain properties. As an example, the SAFER cipher can be pointed out, in which the substitutions are defined by modulo exponentiation and discrete logarithm modulo a prime number operations. Really, such operations are efficient as cryptographic primitive. Therefore, designers believe the substitutions defined by them contribute significantly to the security of the cipher. Another example is the AES algorithm, in which the substitution operation is defined by an algebraic operation over the finite field $GF(2^8)$. In many block ciphers, S-box operations are the main primitive introducing confusion in the encryption process. To link comparatively small data subblocks transformed with S-box operations, these are combined with some simple linear transformations performed over the whole transformed data block, for example, bit permutation operations.

In some designs of block ciphers, table-lookup substitutions are not used. In such cases, confusion is introduced by some nonlinear operations performed on comparatively large data subblocks. Examples are IDEA (a 128-bit block cipher) and RC5 (a block cipher with a parameterized size of the input data block) algorithms. In IDEA, the modulo $2^{16} + 1$ multiplication is used. In the RC5 cipher, the data-dependent rotations (DDR) combined with the modulo 2^{32} addition and subtraction operations are used as the main cryptographic primitive. In the 64-bit version of RC5, the input data block is divided into two 32-bit subblocks, A and B. The DDR is defined as a rotation operation over subblock A, the amount of which is defined by 5 least significant bits of subblock B ($A \leftarrow A <<< B$, where $<<< B$ denotes rotation of B bits to the left; note that only 5 bits of B specify the rotation amount, because the rotation is performed over a 32-bit binary vector). The DDR is also performed by rotating subblock B: $B \leftarrow B <<< A$. The DDRs have been used for the first time by Becker [2] and later by Madryga [27] and Rivest [56]. After Rivest's proposal of the RC5 cipher based on extensive use of DDR, this cryptographic primitive gained recognition. Despite its extreme simplicity, the RC5 cipher has proved to be secure against linear and differential cryptanalysis. It has been shown that mixed use of DDRs with some other simple operations is an effective way to thwart linear cryptanalysis; therefore, DDRs were later used in RC6 [57] and MARS [1] ciphers.

We will show subsequently that DDRs are linear cryptographic primitives, but because of them, after sufficient number of the RC5 iterated algorithm rounds, the modulo 2^{32} additions and subtractions define a good nonlinearity of the resultant transformation. Besides DDRs, different other types of *controlled operations* can be defined and used in the block cipher design. Their important features are (1) their type and (2) the number of different variants from which the current modification is chosen to transform a data subblock. The second characteristic determines how many bits of some other data subblock (called *controlling data subblock*) are significant while performing the controlled operation on the current n-bit data subblock. For a controlled rotation operation, only n modifications are possible. Despite such a small number of modifications, this controlled operation appears to be an efficient cryptographic primitive. We can suppose that controlled operations with an essentially greater number of modifications (for example, 2^n to 2^{3n} or more) will prove to be more efficient. New designs of the controlled operations with a very large number of modifications are connected with the use of so-called PNs, which have been widely studied in the field of parallel processing and telephone switching systems [3,4,11,51]. PNs are well suited for cryptographic applications, because they allow one to specify and perform permutations at the same time. A variant of the symmetric cryptosystem based on PNs and Boolean functions is presented in a study by Portz [53]. Another cryptographic application of PNs is presented by the ICE cipher [23]. In the studies mentioned, PNs are used to specify a key-dependent permutation. Therefore, the operations performed with a PN are fixed permutations and represent linear operations performed on data bit strings; the permutations are, however, unknown to the cryptanalyst. Such use of PNs in the cipher design has been shown [6,58] to be not very effective against differential cryptanalysis. In later works [31,38], it has been shown that cryptology can significantly benefit from the use of PNs to perform *data-driven permutations* or *data-dependent permutations* (DDPs) that are performed with PNs, whereas the current permutation is specified by a data subblock, which is how the DDRs are performed, too. It is easy to construct fast PNs that make it possible to specify all possible permutations on an n-bit data subblock, the number of which is $n! >> 2^n$ for $n > 8$ values. If both the transformed data subblock and the controlling one have the same length ($n = 32$ bits or $n = 64$ bits), then one can easily design a controlled permutation box (CPB) that implements a unique permutation for each possible values of the controlling subblock. This means that the output value of the DDP operation depends on the whole data block. In Chapter 2, different designs of the DDP boxes and DDP-based ciphers are considered in detail.

Some studies [39–41,43–46] propose an approach that advances the DDP primitive. It consists in the use of simple *controlled substitution–permutation networks* (CSPNs) in the data-driven cipher synthesis, instead of PNs. It will be shown subsequently that data-driven operations (DDOs) performed with CSPNs have significantly better properties than the DDP operations. Therefore, CSPNs better suit the designing of fast-hardware-compatible ciphers. The first application of CSPNs relates to their use in key-dependent operations [18], which is significantly less efficient than their role in implementing DDOs.

Another interesting primitive to perform data-driven operations is the so-called *controlled two-place operation* (CTPO). The term CTPO [16] denotes some operation $Y = \mathbf{F}_V(X, A)$ above two operands $\{X, A\} \in \{0, 1\}^n$, which depends on controlling vector $V \in \{0, 1\}^m$ $(m \geq n)$, where $\{0, 1\}^s$ is a set of s-dimensional binary vectors $U = (u_1, u_2, \ldots, u_s)$, $\forall\, i \in \{1, \ldots, s\}\ u_i \in \{0, 1\}$. The simplest variant of such operations is represented by a controlled adder shown in Figure 1.6, which makes it possible to specify 2^n different modifications of the $Y = X *_V A$ addition operation (where V is a controlling vector value), including the bitwise addition modulo 2 ("$*_V$" = XOR) for $V = (0, 0, 0, \ldots, 0)$ and the addition modulo 2^n ("$*_V$" = "$+_n$") for $V = (1, 1, 1, \ldots, 1, 0)$ and $e = 0$ as particular cases.

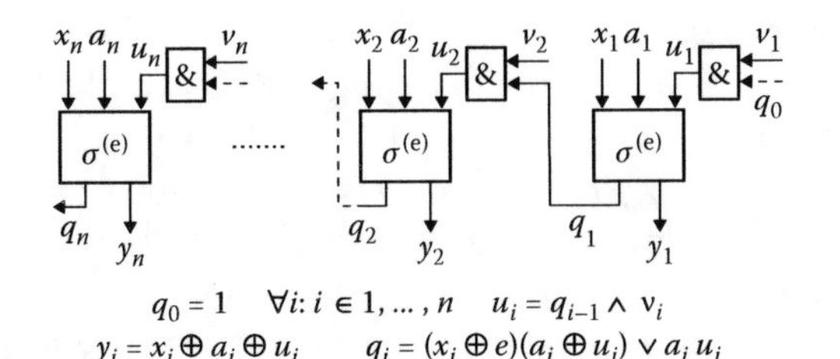

Figure 1.6 Structure of the controlled adder (e = 0 — "addition", e = 1 — "subtraction").

Different designs of the CTPOs are presented in a work by Izotov et al. [16]. A later study [17] describes CTPOs as cellular operations, i.e., a particular case of cellular transformation that comprises cellular operations and cellular automata. The study [17] introduces different linear and nonlinear designs of CTPOs. The CTPO performs bijective or regular transformations, making it possible to modify their nonlinearity and error propagation property (avalanche effect). The CTPO is well suited to define linear and nonlinear transformations that can be implemented with cheap hardware. In some of the block ciphers based on DDP (see Chapter 2), several simple nonlinear CTPOs are used as auxiliary primitives.

The most general type of the controlled operations are controlled table-lookup operations (unfortunately, performing such operations over n-bit data subblocks, where $n \geq 32$ bits, requires extremely large storage, as in the case of table-lookup substitutions). A particular variant of this primitive comprises controlled table-lookup substitutions, which are implemented as follows.

Software implementation of an $n \times n$ substitution uses a table containing two number lines:

$$\begin{pmatrix} 0 & 1 & \dots & 2^n - 1 \\ y_0 & y_1 & \dots & y_{2^n-1} \end{pmatrix} \tag{1.10}$$

where each number represents an n-bit binary vector and the bottom line contains all possible values of the n-bit binary numbers; i.e., it is a permutation of the top line. This permutation defines the specific kind of substitution operation. To perform an S-box operation on some n-bit vector x with this table, we interpret this vector as a binary number of value x and take the value y_x as the output of the substitution operation. Let $S(x)$ denote a substitution table. Suppose we have defined 2^m different substitution tables

$$S_0(x), S_1(x), \dots, S_{2^m-1}(x) \tag{1.11}$$

Then we define the controlled S-box operation: $y = S_V(x)$, where V is controlling m-bit binary vector. The implementation of this controlled operation requires storage of 2^m substitution tables, which can be defined using different security criteria. For example, in the case $n = m$, the following criterion can be justified: for each fixed input value x_0 the set of substitution tables $S_V(x)$, where $V = 1, 2, \ldots, 2^m - 1$, should define the function $y = F_{x_0}(V)$, which is a substitution $y = S'_{x_0}(V)$. The $S_V(x)$ operation is efficient and fast, but it needs expensive hardware and occupies very large memory space in software implementation. Generally, we can use 2^m different tables of arbitrary type, with which a controlled table-lookup operation can be defined, as in the case of controlled substitutions.

For fast software encryption, some studies [32–34] have introduced block cipher design based on data-dependent subkey selection. In such data-driven operations, the encryption process is performed with simple two-place operations (XOR, modulo 2^{32} addition, and modulo 2^{32} subtraction) performed on a transformed data subblock and a subkey selected depending on the value of another subblock that is a current controlling data subblock. Some authors [35,36] advance the block cipher design based on data-dependent subkey selection. A study by Moldovyan et al. [37] presents a cipher with a key-dependent algorithm with provably inequivalent encryption modifications, which is also based on data-dependent subkey selection. This primitive represents a trade-off between the efficiency of table-lookup operations and the required memory.

Thus, at present, we have the following general types of primitives for designing ciphers based on data-driven operations:

- Controlled permutations including controlled rotations as a particular variant
- Controlled operations implemented with CSPN (we will call them controlled operational substitutions [COSes])
- Controlled two-place operations
- Controlled table-lookup operations, including controlled substitutions
- Data-dependent subkey selection

In the following chapters we will consider mainly designs of the first two primitives and their use in the data-driven cipher synthesis, because these primitives well suit efficient and fast hardware. We will also consider encryption algorithms based on data-dependent subkey selection, implemented in fast software. Such algorithms are interesting for applications in communication devices with embedded processors.

1.1.6 Construction Scheme Variants of Iterated Ciphers Based on Data-Dependent Operations

Considering iterated structures of the ciphers presented in the following text, we assume that some simple key scheduling is used; i.e., subkeys of the secret key are directly used in encryption process. The proposed cryptoschemes describe several types of iterated transformations using controlled operations, which are conventionally denoted as **F**, $\mathbf{F}^{-1}$, $\mathbf{F_i}$, and **S** boxes. These operations represent basic primitives. The auxiliary operations are the XOR and extension box **E**, which mean very cheap hardware (simple wiring). We need the **E** box operation, because in the usually used controlled operations implemented with PNs and CSPNs the controlling vector is longer than the transformed binary vector. The **F** and $\mathbf{F}^{-1}$ boxes perform mutually inverse controlled operations; i.e., at the same value of the controlling vector V, they perform mutually inverse transformations. The box $\mathbf{F_i}$ represents controlled involution; i.e., for each of possible value of the controlling vector V, it implements an operation that is an involution. Thus, for all possible values, V and X, of the hypothetical controlled operations performed, the following expression holds:

$$X = \mathbf{F}^{-1(V)}(\mathbf{F}^{(V)}(X)),\ X = \mathbf{F}^{(V)}(\mathbf{F}^{-1(V)}(X)),\quad \text{and}\quad X = \mathbf{F_i}^{(V)}(\mathbf{F_i}^{(V)}(X)) \tag{1.12}$$

Cryptoscheme, shown in Figure 1.7*a*, can be described using the following general notation:

$$(A_j, B_j) = \mathrm{Crypt}(A_{j-1}, B_{j-1}, K_j, Q_j) \tag{1.13}$$

where (A_{j-1}, B_{j-1}) and (A_j, B_j) are input and output data blocks, represented as the concatenation of two subblocks of the same size, and (K_j, Q_j) is the jth-round key, represented as the concatenation of subkeys K_j and Q_j. Selecting this design scheme, one can construct different iterated ciphers, using different pairs of the controlled operations **F** and $\mathbf{F}^{-1}$. The number of the encryption rounds, R, should be assigned

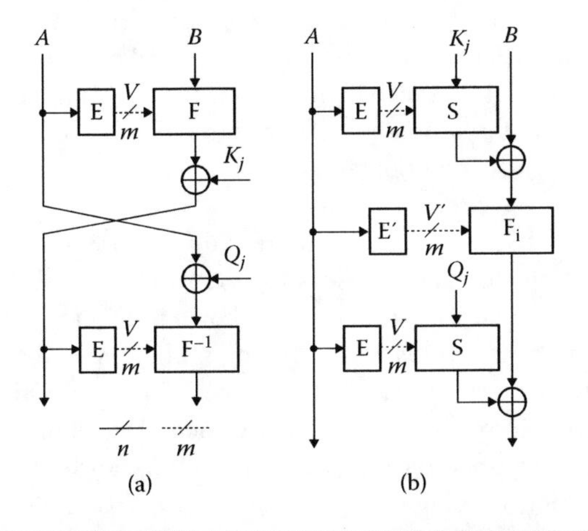

Figure 1.7 Encryption round with mutually inverse controlled operations (*a*) and with controlled involution (*b*).

depending on the concrete type of the controlled operations performed. The structure of the round transformation is composed so that the changing encryption process for decryption is realized as a simple modification of the key scheduling. Note that the round transformation **Crypt** is not an involution and the jth decryption round is defined by transposing the subkeys K_j and Q_j:

$$\mathbf{Crypt}(A_j, B_j, Q_j, K_j) = (A_j', B_j') \tag{1.14}$$

$$\mathbf{Crypt}(A_j', B_j', K_j, Q_j) = (A_j, B_j) \tag{1.15}$$

Let (K_j, Q_j) be the jth encryption-round key and (K_j', Q_j') be the jth decryption-round key. Then the decryption process is correct if the key scheduling is such that the condition $(K_j, Q_j) = (Q'_{R-j+1}, K'_{R-j+1})$ holds.

The round transformation in Figure 1.7*b* is also not an involution, and its inverse is defined by transposing the subkeys K_j and Q_j. Because of the use of controlled involution, it is not necessary to perform two mutually inverse controlled operations. Potentially, this design element makes it possible to simplify the transformation-round structure. The peculiarity of the scheme is the use of data-dependent transformation of subkeys K_j and Q_j with the COS box **S**. Transformation of the subkeys of the round, depending on transformed data, can be called *internal key scheduling.* The advantage of the scheme in Figure 1.7*b* is its ability to perform in parallel the transformation of the right data subblock with the $\mathbf{F}_\mathbf{i}$ operation and the transformation of the Q_j subkey with the **S** box. Thus, one of two operations performed on subkeys introduces no delay in the encryption-round process. The decryption is correct if the key scheduling is such that the condition $(Q'_{R-j+1}, K'_{R-j+1}) = (K_j, Q_j)$ holds, as in the case of Figure 1.7*a*. The general iterated structure of the data ciphering process using two considered round transformations is presented in Figure 1.8*a*. It is possible to have iterated ciphers in which the round structure is not symmetric. In such ciphers, after the Rth round of transformation, some simple final transformation is performed (see Figure 1.8*b*), introducing symmetry, due to which the encryption and decryption processes differ only in the key scheduling.

Figures 1.9*a* and 1.9*b* show encryption-round structures that require the use of a final transformation that adds subkeys to data subblocks with the XOR operation. The first scheme is derived from the one shown in Figure 1.7*a*. In this scheme, the round transformation begins with XORing subkeys K_j and Q_j with data subblocks A and B, respectively. Then, the **F** box operation, transposition of subblocks A and B, and the $\mathbf{F}^{-1}$ box operation are performed. The whole ciphering process represents an operation of performing consecutively R transformation rounds, and a final transformation (see Figure 1.9*c*) defined as XORing data subblocks with subkeys S and T:

$$(A', B') = (A_R \oplus S, B_R \oplus T) \tag{1.16}$$

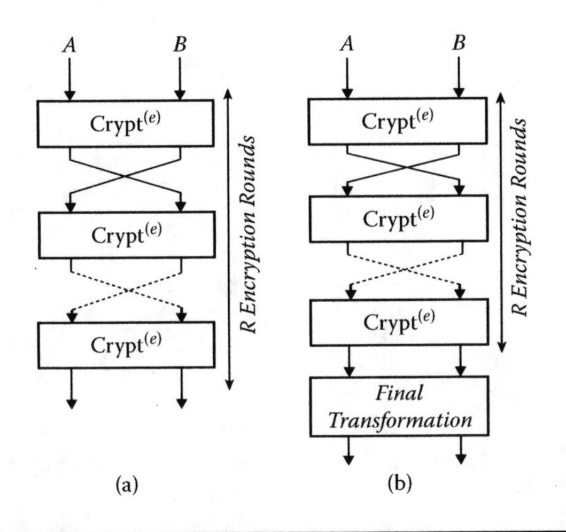

Figure 1.8 General structure of the iterated ciphering process.

The simpler round structure of Figure 1.9*b* is due to the using of controlled involution $\mathbf{F}_i$. To define the correct decryption, the key scheduling should be such that the following relations hold:

$$K'_1 = T;\ K'_j = Q_{R-j+2} \quad \text{for} \quad j = 2, 3, \ldots, R;\ S' = Q_1 \tag{1.17}$$

$$Q'_1 = S;\ Q'_j = K_{R-j+2} \quad \text{for} \quad j = 2, 3, \ldots, R;\ T' = K_1 \tag{1.18}$$

Some advancing of the schemes shown in Figure 1.7*b* can be obtained by adding the transformation of the left data subblock with a key-dependent controlled involution. This additional operation does not introduce any time delay during hardware implementation, because it is performed in parallel with transforming the K_j subkey, with the **S**-box operation. After this modification, the indicated round transformations get the structures presented in Figure 1.10*a*. The modified scheme includes two controlled involutions, $\mathbf{F}'_i$ and $\mathbf{F}_i$, but only one of them, namely $\mathbf{F}_i$, is data driven. The peculiarity of the scheme shown in Figure 1.10*a* is the XORing of input and output of the $\mathbf{F}'_i$ operation to form the input value of the $\mathbf{E}'$ operation. The structure of another attractive round is presented in Figure 1.10*b*, in which two data subblocks are transformed simultaneously. In both cryptoschemes, the $\mathbf{F}'_i$ operation is controlled with the subkey of the round, U_j, which is independent of subkeys K_j and Q_j or can be defined as $U_j = K_j \oplus Q_j$. Instead of the key-dependent operation $\mathbf{F}'_i$, we can use a cascade of substitutions having size 4 × 4 or 8 × 8, all of which are involutions. In this case, we need no additional round subkey. In the case of the round transformation presented in Figure 1.10*a*, the correctness of the decryption process is defined by the following formulas:

$$K'_j = Q_{R-j+1};\ Q'_j = K_{R-j+1};\ U'_j = U_{R-j+1}, \quad \text{where} \quad j = 1, 2, \ldots, R \tag{1.19}$$

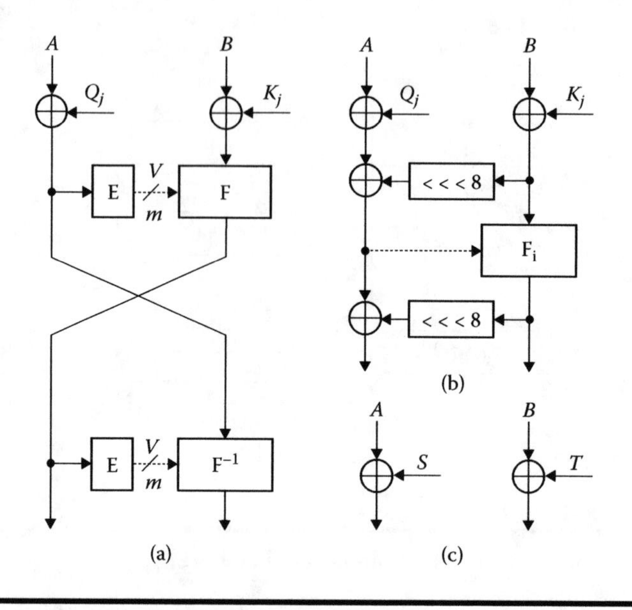

Figure 1.9 Nonsymmetric round transformations (*a*, *b*) and final transformation (*c*).

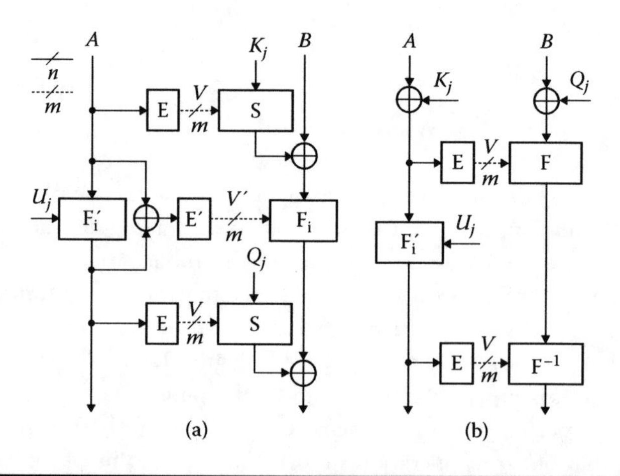

Figure 1.10 Symmetric (*a*) and nonsymmetric (*b*) round transformations with simultaneous transformation of both data subblocks.

We have considered only some of characteristic structures of the encryption round, including the DDOs. Block ciphers presented in the literature illustrate some other round transformation designs [41–47].

In the case of the round transformation shown in Figure 1.10*b*, the final transformation is to be effected after the *R*th round. In the last case, the correctness of the decryption process is defined by formulas (1.17) and (1.18), as well as the formula $U'_j = U_{R-j+1}$ for $j = 1$ to R.

Chapter 2

Permutation Networks as Primitive of Data-Driven Ciphers

This chapter describes permutation networks (PNs) as primitive for implementing data-dependent permutations (DDPs). The notion of the order of the PN is introduced, and different topologies of the PNs are considered. Linear and differential properties of DDPs are investigated. Several DDP-based ciphers are described, and estimation of their security against linear and differential analysis is presented.

2.1 Design of the Permutation Networks

PNs have been invented to perform fast controlled interconnections, while solving different noncryptographic problems [3,4,11,68]. The first cryptographic applications of PNs relate to their use as key-dependent operations [23,53]. However, detailed cryptanalysis has shown that ciphers based on key-dependent permutations are not able to survive in the speed competition with other block ciphers [6,58]. In some studies [14,31,38], PNs have been proposed to implement DDPs in block ciphers. Security analysis of DDP-based ciphers [22,42,47,48] has shown that PNs are more useful in this method than in performing key-dependent permutations. The principal advantage of the DDPs against the key-dependent permutations that are fixed is the variability while encrypting different data blocks. The usage of PNs to perform variable permutations has stimulated a new interest the use of the PNs in applied cryptography. Initially, the DDP has been proposed as primitive for designing fast ciphers suitable for cheap hardware implementation. Hardware implementation estimates [49,64,65] of these ciphers confirmed that they provide fast and cheap cryptographic hardware. A critical argument against using the DDP as a cryptographic primitive is low software performance. In reply to this argument, it has been proposed that controlled permutation (CP) instructions (CPI) be embedded in general-purpose processors [24–26,48] to increase simultaneously the software encryption speed of DDP-based algorithms and the performance of non-cryptographic algorithms, including fixed bit permutations of arbitrary types. The embedding of CPI in the standard set of elementary commands is a very attractive way to update general-purpose microprocessors. Indeed, the CPI is highly desired to decrease the number of required instructions to perform bit permutation operations on conventional processors. In Chapter 4, different architectures of the CPI oriented to cryptographic and general-purpose applications are considered. Thus, we can suppose that the DDP primitive will be available for software implementation on future processors.

In the following text, considering the cryptographic applications of PNs, we shall call them *controlled permutation boxes* (CPBs). The term CPB underlines its use in performing CPs on binary vectors. Let $\mathbf{P}_{n/m}$ denote a CPB with n-bit input and m-bit control input. CPB are usually constructed using elementary switching elements $\mathbf{P}_{2/1}$ as elementary building blocks. Each of the $\mathbf{P}_{2/1}$ boxes is controlled with one bit, v, and passes ($v = 0$) or transposes ($v = 1$) two input bits x_1 and x_2. In other words, the elementary switching element forms at its output two-bit value (y_1, y_2), where $y_1 = y_{1+v}$ and $y_2 = x_{2-v}$. A $\mathbf{P}_{n/m}$ box has a layered topology and can be represented (see Figure 2.1) as a superposition,

$$\mathbf{P}^{(V)}_{n/m} = \mathbf{L}^{(V1)\circ}\pi_1\, \mathbf{L}^{(V2)}\, \pi_2 \ldots \pi_{s-1}\, \mathbf{L}^{(V_s)} \tag{2.1}$$

where $\mathbf{L}$ is an active layer composed of $n/2$ switching elements; $V_1, V_2, \ldots, V_s$ are controlling vectors of the active layers from 1 to s; $\pi_1, \pi_2, \ldots, \pi_{s-1}$ are fixed permutations; and $V = (V_1, \ldots, V_s)$ is the controlling vector of the $\mathbf{P}_{n/m}$ box. Controlled permutations performed with the $\mathbf{P}_{n/m}$ box can be characterized using an ordered set of the modifications $\{\Pi_0, \Pi_1, \Pi_2, \ldots, \Pi_{2m-1}\}$, where each modification Π_i, $i = 0, 1, \ldots, 2^m - 1$, is a fixed permutation of some set of n bits. Permutations Π_i (or $\mathbf{P}_i$) will be called CP modifications. The execution of the CP operation $\mathbf{P}^{(V)}_{n/m}(X)$ consists in performing the permutation Π_V on X: $Y = \mathbf{P}^{(V)}_{n/m}(X) = \Pi_V(X)$. The following two definitions [38] introduce some notions useful in the block cipher design:

Definition 2.1. *Suppose a CPB $\mathbf{P}_{n/m}$ is given. A CPB $\mathbf{P}^{-1}_{n/m}$ is called the inverse of $\mathbf{P}_{n/m}$ box if, for all V, the corresponding CP modifications $\mathbf{P}_V$ and $\mathbf{P}^{-1}_V$ are mutual inverses.*

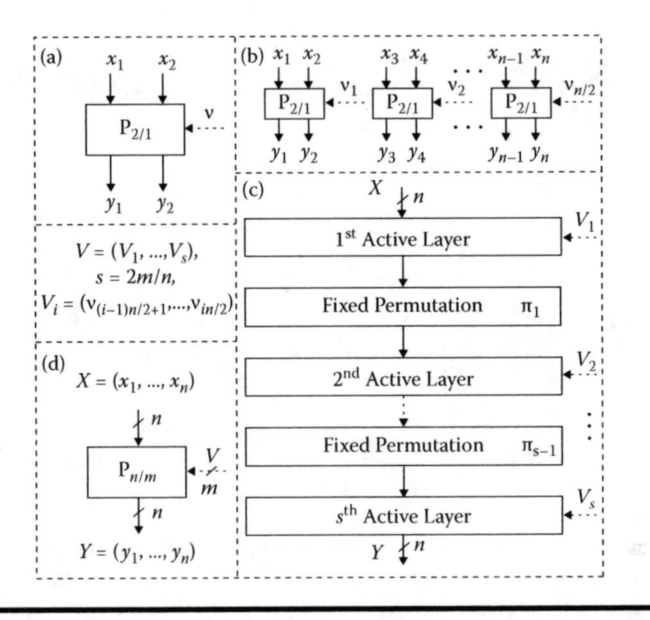

Figure 2.1 Notation of the $P_{2/1}$ (*a*) and $P_{n/m}$ boxes (*b*), structure of one active layer (*c*), and general structure of the layered CP boxes (*d*).

Definition 2.2. *Suppose that for arbitrary $h \leq n$ input bits $x_{\alpha 1}, x_{\alpha 2}, ..., x_{\alpha h}$ and arbitrary h output bits $y_{\beta 1}, y_{\beta 2}, ..., y_{\beta h}$ there is at least one value of the controlling vector, V, that specifies a CP-box permutation, $\mathbf{P}_V$, moving $x_{\alpha i}$ to $y_{\beta i}$ for all $i = 1, 2, ..., h$. Such $\mathbf{P}_{n/m}$ box is called a CP box of order h.*

One active layer can be considered as the single-layer CP box $\mathbf{L}_n$ having n-bit input and representing a cascade of $n/2$ elementary switching elements. It is evident that $\mathbf{P}_{1/2} = \mathbf{P}^{-1}_{1/2}$; therefore, $\mathbf{L}^{-1}_n = \mathbf{L}_n$. An arbitrary CPB operation is invertible with another CPB. Inverse transformation can be constructed by swapping input and output of the given CP box. Let $\mathbf{P}^{-1}_{n/m}$ be the formal notation of the inverse transformation corresponding to the direct transformation $\mathbf{P}_{n/m}$. Thus, the inverse CP box has the following structure:

$$\mathbf{P}^{-1}_{n/m} = \mathbf{L}_n^{(V_s)} \circ \pi^{-1}_{s-1} \circ \mathbf{L}_n^{(V_{s-1})} \circ \pi^{-1}_{s-2} \circ \ldots \circ \pi^{-1}_1 \circ \mathbf{L}_n^{(V1)} \quad (2.2)$$

The components $V_1, V_2, \ldots, V_s$ compose the controlling vector V of both $\mathbf{P}_{n/m}$ and $\mathbf{P}^{-1}_{n/m}$ boxes: $V = (V_1, V_2, \ldots, V_s)$. Let the switching elements in the $\mathbf{P}_{n/m}$ (or $\mathbf{P}^{-1}_{n/m}$) boxes be consecutively enumerated from left to right and from top to bottom (or from bottom to top). Then the ith bit of the V vector controls the ith elementary box in $\mathbf{P}_{n/m}$ and $\mathbf{P}^{-1}_{n/m}$ boxes.

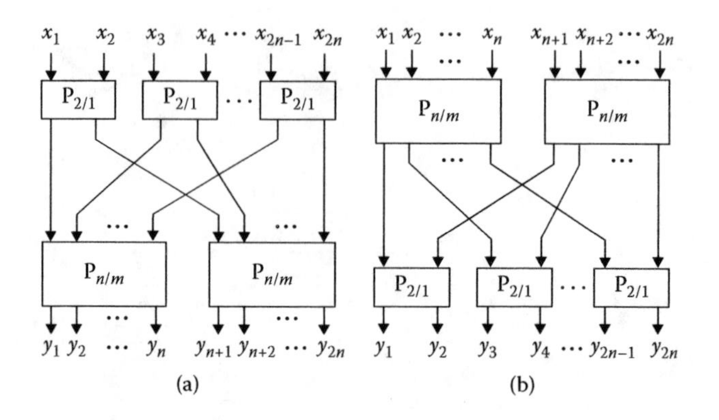

Figure 2.2 The recursive construction of the first (*a*) and second (*b*) types: structure of the boxes $P_{2n/2m+n}$ (*a*) and $P^{-1}_{2n/2m+n}$ (*b*).

Selecting different variants of fixed permutations, one can define different types of CP boxes for given values n and m ($m = ns/2$). Let us consider the most interesting case $n = 2^k$, where k is a natural number. Figure 2.2*a* presents a variant of the recursive construction of some class of CP boxes with a layered structure. This construction is called the recursive construction of the first type. It allows one to compose the box $\mathbf{P}_{2n/2m+n}$ of the first order from two $P_{n/m}$ boxes of order $1 \le h \le n$ and one additional layer with n boxes $\mathbf{P}_{2/1}$. Starting with $n = 2$ and $m = 1$, one can consecutively construct the following CP boxes of the first order: $\mathbf{P}_{4/4}$, $\mathbf{P}_{8/12}$, $\mathbf{P}_{16/32}$, $\mathbf{P}_{32/80}$, $\mathbf{P}_{64/192}$, and so on. In an analogous way, using the recursive construction of the second type (see Figure 2.2*b*), one can compose the CP box $\mathbf{P}^{-1}_{2n/2m+n}$ from two $\mathbf{P}^{-1}_{n/m}$ boxes and n boxes $\mathbf{P}_{2/1}$. It is easy to see that for $n = 2^k$ the minimum number of layers in layered CP boxes of the first order is $s = \log_2 n$.

Another example of recursive structure is presented in Figure 2.3*a*, in which it is shown how a CP box $\mathbf{P}_{2n/(2m+2n)}$ of order $2h$ is constructed from two $\mathbf{P}_{n/m}$ boxes of order $\ge h$ and two additional $\mathbf{P}_{2/1}$ layers [50]. Figure 2.3*b* corresponds to a recursive construction of the box $\mathbf{P}_{4n/(4m+8n)}$ of the order $4h$ using the boxes $P_{n/m}$ of order $\ge h$. Thus, the recursive construction defined by the mechanism shown in Figure 2.3*a* allows doubling of the order of internal boxes $\mathbf{P}_{n/m}$ (or $\mathbf{P}^{-1}_{n/m}$). In some more general applications, two internal boxes $\mathbf{P}_{n/m}$ and $\mathbf{P}'_{n/m}$ have different orders $h \ne h'$. In this case, the statement on doubling of the order with a third-type recursive construction has the following form.

Statement 2.1. *Each step of the third-type recursive construction doubles the minimum order* $\min\{h, h'\}$ *of the internal boxes* $\mathbf{P}_{n/m}$ *and* $\mathbf{P}'_{n/m}$.

This statement is sufficiently evident. Suppose we have $2\cdot\min\{h, h'\}$ arbitrary indicated input bits. Then, defining controlling bits of the top active layer, we can move half of the indicated bits (i.e., $\min\{h, h'\}$ bits) to the input of the $\mathbf{P}_{n/m}$ box and

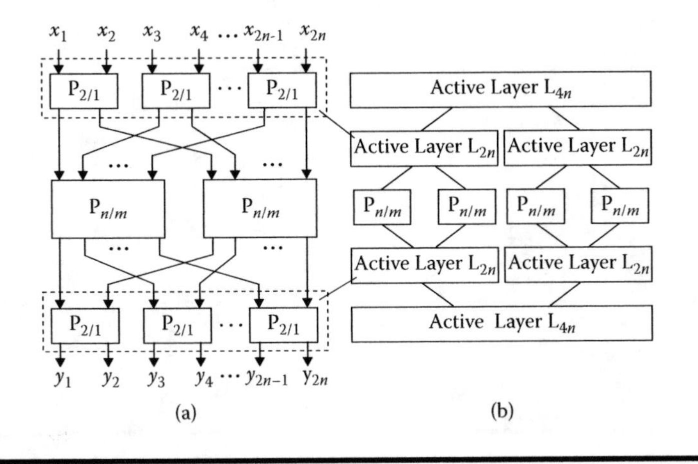

Figure 2.3 The third-type recursive construction: structure of the CP boxes $\mathbf{P}_{2n/2m+2n}$ (*a*) and $\mathbf{P}_{4n/4m+8n}$ (*b*).

move the other half (i.e., min$\{h, h'\}$ bits) to the input of the $\mathbf{P}'_{n/m}$ box. Because the order of each of the internal boxes is equal to or greater than min$\{h, h'\}$, there exist values of the controlling vectors of boxes $\mathbf{P}_{n/m}$ and $\mathbf{P}'_{n/m}$ such that each of the indicated bits is moved to any indicated switching element of the bottom active layer. Therefore, the indicated bits can be moved arbitrarily to $2\cdot\min\{h, h'\}$ digits of the output of box $\mathbf{P}_{2n/(2m+2n)}$. This is a simple, but not formal, justification (the formal proof is long and annoying).

Thus, the resultant box $\mathbf{P}_{2n/(2m+2n)}$ has the order $h'' = 2\cdot\min\{h, h'\}$. Starting with $n = 2$ and $m = 1$ and using the recursive mechanism of the third type, presented in Figure 2.3, one can consecutively construct the following CP boxes of the maximum order ($h = n$): $\mathbf{P}_{4/6}$, $\mathbf{P}_{8/20}$, $\mathbf{P}_{16/56}$, $\mathbf{P}_{32/144}$, $\mathbf{P}_{64/352}$, and so on. Note that in the third-type recursive construction, we add two active layers of the same size at each step. At the *i*th step, two layers $\mathbf{L}_{2^{i+1}}$ (upper and lower ones) are added. The number of layers that are required to construct the maximum-order CP box is $s = 2\log_2 n - 1$. It is worth noting that these maximum-order boxes have mirror symmetry topology. Such symmetry is often used in designing DDP primitives of iterative block ciphers. In general, boxes of the order $h < n$ are not symmetric, but for particular values of n and h ($h < n$) it is possible to construct $\mathbf{P}_{n/m}$ boxes having mirror symmetry topology.

One can show that it is possible to construct CP boxes of the order $h = 1, 2, \ldots, n/4$ using $s = \log_2 nh$ active layers. Indeed, in some maximum-order boxes, $\mathbf{P}_{n/m}$, having the third-type recursive topology, we can delete all lower layers, $\mathbf{L}_4$, corresponding to the first step of the recursive construction. This gives the CP box $\mathbf{P}_{n/m'}$ of the order $h = n/4$. Deleting all lower layers, $\mathbf{L}_8$, corresponding to the second step of the recursive construction, we get a CP box, $\mathbf{P}_{n/m''}$, of the order $h = n/8$. Similarly, we can get CP boxes with n-bit input of the order $h = n/8, n/16, \ldots, 2, 1$. At the last of such

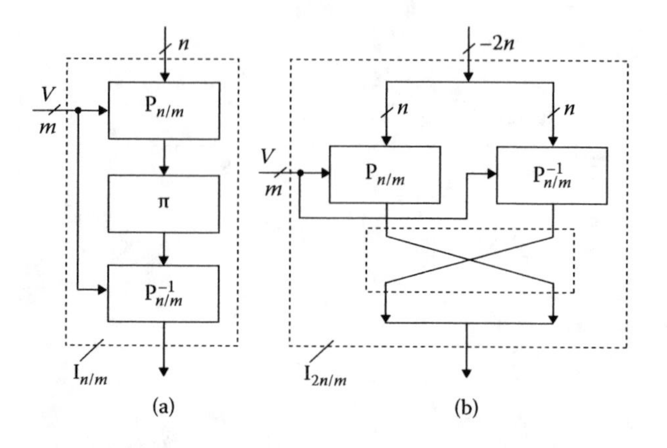

Figure 2.4 Two design variants of the controlled permutational involutions: serial (*a*) and parallel (*b*) structures.

steps, we get the topology corresponding to the first recursive type. Note that for a given value of n in CP boxes of the orders $h_1 = n/4$ and $h_2 = n$, the number of active layers are s_1 and $s_2 = s_1 + 1$. Therefore, construction of CP boxes of the order $h = n/2$ is not of practical interest; for applications of CP boxes as cryptographic primitives, their order could be $h = 1, 2, \ldots, n/4, n$.

Using recursive constructions of the three types considered previously, we get a topology with mirror symmetry only for CPBs of maximum order. The case of symmetric topology is of special interest, because in this case mutually inverse CP boxes $\mathbf{P}_{n/m}$ and $\mathbf{P}^{-1}_{n/m}$ differ only in the distribution of controlling bits, i.e., the box $\mathbf{P}_{n/m}$ implements two mutually inverse permutations for the following two values of the controlling vector: $V = (V_1, V_2, \ldots, V_{s-1}, V_s)$ and $V = (V_s, V_{s-1}, \ldots, V_2, V_1)$; i.e., transposing the binary vectors V_{s-i+1} and V_i for $i = 1, 2, \ldots, [s/2]$, we get the inverse CPB $\mathbf{P}^{-1}_{n/m}$ from the $\mathbf{P}_{n/m}$ box. This property is used in Chapter 4, in which different variants of switchable controlled operations are introduced. For particular values of n and $h < n$, it is also possible to construct CPB $\mathbf{P}_{n/m}$ with symmetric topology. For example, it is easy to design symmetric topology for: (1) four-layer CPB $\mathbf{P}_{16/32}$ of the first order; (2) six-layer CPB $\mathbf{P}_{32/96}$ of the second order, and (3) six-layer CPB $\mathbf{P}_{64/192}$ of the first order.

In some cases, we need to implement DDP operations all permutation modifications of which are involutions. Such operations can be implemented with the CP boxes of either (1) serial or (2) parallel structure. The first structure is shown in Figure 2.4*a*. It represents the superposition $\mathbf{P}_{n/m}{}^{\circ}\pi^{\circ}\mathbf{P}^{-1}_{n/m}$, where π is a fixed permutation with required structure of cycles; the $\mathbf{P}_{n/m}$ and $\mathbf{P}^{-1}_{n/m}$ boxes are controlled with the same controlling vector. For the arbitrarily fixed value V, the CPB $\mathbf{I}_{n/m} = \mathbf{P}_{n/m}{}^{\circ}\pi^{\circ}\mathbf{P}^{-1}_{n/m}$ implements the fixed permutation Π_V having the same cyclic structure as the π fixed permutation. In general, for different values of V, different modifications, Π_V, are implemented, all of them having the same cyclic structure. If π is a permutational involution, then all modifications, $\Pi_{V,}$ are involutions. Indeed, for some fixed value, V, we have

$$\mathbf{I}_{n/m}{}^{\circ}\mathbf{I}_{n/m} = (\mathbf{P}_{n/m}{}^{\circ}\pi^{\circ}\mathbf{P}^{-1}_{n/m})^{\circ}(\mathbf{P}_{n/m}{}^{\circ}\pi^{\circ}\mathbf{P}^{-1}_{n/m}) = \mathbf{P}_{n/m}{}^{\circ}\pi^{\circ}\mathbf{P}^{-1}_{n/m}{}^{\circ}\mathbf{P}_{n/m}{}^{\circ}\pi^{\circ}\mathbf{P}^{-1}_{n/m} =$$

$$\mathbf{P}_{n/m}{}^{\circ}\pi^{\circ}\pi^{\circ}\mathbf{P}^{-1}_{n/m} = \mathbf{P}_{n/m}{}^{\circ}\mathbf{P}^{-1}_{n/m} = \mathbf{O} \tag{2.3}$$

where $\mathbf{O}$ is an identical permutation; i.e., the considered superposition represents a controlled permutational involution. If π is a one-cycle permutation, then all modifications, $\Pi_{V,}$ are one-cycle fixed permutations, and in general Π_V are different for different V. Because of its serial structure, this type of the controlled involution stipulates the doubling of the time delay of the $\mathbf{I}_{n/m}$ box operation in comparison with the $\mathbf{P}_{n/m}$ box. To get smaller time delay values, the parallel structure of the controlled involutions is used, which is presented in Figure 2.4*b*. In the second structure, two mutually inverse CP boxes $\mathbf{P}_{n/m}$ and $\mathbf{P}^{-1}_{n/m,}$ compose a cascade, $\mathbf{P}_{n/m} \,||\, \mathbf{P}^{-1}_{n/m,}$ followed by the transposition operation $\mathbf{T}$. Thus, we have the following controlled permutational involution: $\mathbf{I}_{2n/m} = (\mathbf{P}_{n/m} \,||\, \mathbf{P}^{-1}_{n/m})^{\circ}\mathbf{T}$. Indeed, for the arbitrarily fixed value V we have

$$\mathbf{I}_{2n/m}{}^{\circ}\mathbf{I}_{2n/m} = ((\mathbf{P}_{n/m} \,||\, \mathbf{P}^{-1}_{n/m})^{\circ}\mathbf{T})^{\circ}((\mathbf{P}_{n/m} \,||\, \mathbf{P}^{-1}_{n/m})^{\circ}\mathbf{T}) = (\mathbf{T}^{\circ}(\mathbf{P}^{-1}_{n/m} \,||\, \mathbf{P}_{n/m}))^{\circ}((\mathbf{P}_{n/m} \,||\, \mathbf{P}^{-1}_{n/m})^{\circ}\mathbf{T}) =$$

$$\mathbf{T}^{\circ}((\mathbf{P}^{-1}_{n/m} \,||\, \mathbf{P}_{n/m})^{\circ}(\mathbf{P}_{n/m} \,||\, \mathbf{P}^{-1}_{n/m}))^{\circ}\mathbf{T} = \mathbf{T}^{\circ}\mathbf{T} = \mathbf{O} \tag{2.4}$$

The parallel design topology of the controlled involutions will be used subsequently while composing different types of cryptographic primitives, including the so-called switchable data-driven operations (see Chapter 4).

2.2 Linear Characteristics of the Controlled Permutations

Linear characteristics of the DDP operation characterize its contribution to security against linear cryptanalysis. They have been investigated in a recent study [47], some results of which are considered here. The following specific notations are used in this section:

- The parity of $\varphi(U)$, i.e., $\varphi'(U) \overset{def}{=} \varphi(U) \bmod 2$ is denoted by $\varphi'(U)$.
- Let $X \otimes Y$ denote the bitwise AND operation of the two vectors X and Y: $X, Y \in \{0, 1\}^{(s)}$. For $c \in \{0, 1\}$ and $X \in \{0, 1\}^{(s)}$, we define $Y = c \cdot X$, where $y_i = c \cdot x_i$ $\forall$ $i \in \{1, ..., s\}$.
- Let • denote the binary scalar product: $c = A \bullet X = \varphi'(A \otimes X)$, $c \in \{0, 1\}$.
- Let $E_0 = (0, ..., 0)$ and $D_0 = (1, ..., 1)$; i.e., $\varphi(E_0) = 0$ and $\varphi(D_0) = s$, where E_0, $D_0 \in \{0, 1\}^s$.

Let **F**: $\{0, 1\}^r \rightarrow \{0, 1\}^n$ $(r \geq n)$ be a given transformation function, $Y = \mathbf{F}(U)$. Its resistance against linear cryptanalysis (LCA) [28] is determined by the maximum *bias* (or *deviation*) b, which is defined as follows:

$$b = b(\mathbf{M}^U, \mathbf{M}^Y) = \left| \Pr_U(U \bullet \mathbf{M}^U \oplus Y \bullet \mathbf{M}^Y = 0) - \frac{1}{2} \right| \tag{2.5}$$

where U, $\mathbf{M}^U \in \{0, 1\}^r$, $Y \in \{0, 1\}^n$, $\mathbf{M}^Y \in \{0, 1\}^n$. The fixed vectors $\mathbf{M}^U$ and $\mathbf{M}^Y$ are called masks of vectors U and Y, respectively. A mask selects some subset of bits of the considered vector. For example, nonzero bits of masks correspond to positions of the input and output data bits that are taken into account while calculating the b value. Such bits of masks or indicated positions are called *active bits*.

The bias b depends on the selection of the $\mathbf{M}^U$ and $\mathbf{M}^Y$ masks. The triple $(\mathbf{M}^U, \mathbf{M}^Y, b)$ is called *linear characteristic* (LC) of the function **F**. The bias value depends on the selected input and output masks. While performing linear cryptanalysis [28] an attacker can select masks to which the higher bias values correspond (the higher the value of b, the lower the resistance of the **F** function against linear cryptanalysis). Therefore, the resistance of some function against LCA is determined by the maximum value of the bias in the set of all possible LCs. Let **F** be a function in two variables defined as $Y = \mathbf{P}_{n/m}(X, V)$, where $X \in \{0, 1\}^n$ and $V \in \{0,1\}^m$. In this case, we have $U = (X, V)$ and $\mathbf{M}^U = (\mathbf{M}^X, \mathbf{M}^V)$; therefore, equation (2.5) is transformed into

$$b(\mathbf{M}^X, \mathbf{M}^V, \mathbf{M}^Y) = \left| \Pr_{(X,V)}\left(X \bullet \mathbf{M}^X \oplus V \bullet \mathbf{M}^V \oplus Y \bullet \mathbf{M}^Y = 0\right) - \frac{1}{2} \right| \tag{2.6}$$

An input mask can be considered as input vector that is transformed by the $\mathbf{P}_{n/m}$ box. It is sufficiently evident that the bias of some LC of the CP box is equal to zero if $\varphi(\mathbf{M}^X) \neq \varphi(\mathbf{M}^Y)$; therefore, for an LC with nonzero bias, we can conclude that the input mask is transformed into output mask, because in this case $\varphi(\mathbf{M}^X) = \varphi(\mathbf{M}^Y)$. Assuming that the controlling vector is uniformly distributed value, we can consider the probability

$$\Pr\left(\mathbf{M}^Y = \mathrm{P}_{n/m}\left(\mathbf{M}^X\right)\right) = p\left(\mathbf{M}^X \rightarrow \mathbf{M}^Y\right) \tag{2.7}$$

If the CP box is of the order $h \geq 1$, then $p(\mathbf{M}^X \rightarrow \mathbf{M}^Y) \neq 0$ for all masks such that $\varphi(\mathbf{M}^X) = \varphi(\mathbf{M}^Y) \leq h$. The study mentioned previously [47] considers in detail the LC of the CP boxes of the orders $1 \leq h \leq n$ and gives the formal proof of the following statement.

Statement 2.2. *The absolute value of the deviation of LC with a nonzero mask of the controlling vector does not exceed the value of the deviation of LC with a zero mask of V, the last being equal to half of the probability that a given input mask transforms into a given output mask.*

The practical significance of Statement 2.2 consists in the simplification of calculation of bias values for different LCs. The following practical recommendations are derived from this statement:

- To estimate the LC of CP boxes, one need analyze only characteristics with zero mask of the controlling vector V.
- In a certain sense, the calculation of the bias b corresponding to the LC ($\mathbf{M}^X$, $\mathbf{M}^Y$, b) is equivalent to the calculation of the probability $p(\mathbf{M}^X \rightarrow \mathbf{M}^Y)$ that the CP box transforms vector $\mathbf{M}^X$ into vector $\mathbf{M}^Y$. For $\varphi(\mathbf{M}^X) \neq \varphi(\mathbf{M}^Y)$, we have $p(\mathbf{M}^X \rightarrow \mathbf{M}^Y) = 0$ and $b = 0$.
- For CP boxes of the order h, where $1 \leq h \leq n$, for any LC with masks $\mathbf{M}^X$, $\mathbf{M}^Y \notin \{E_0, D_0\}$ we have $p(\mathbf{M}^X \rightarrow \mathbf{M}^Y) \leq 1/n$ and $b \leq 1/2n$.
- The DDP box is a linear cryptographic primitive having a single LC with maximum weight masks $\mathbf{M}^X = \mathbf{M}^Y = (1, \ldots, 1)$. Therefore, while designing cryptoschemes with the use of DDP boxes, other additional auxiliary nonlinear operations should be applied to thwart linear attacks using the maximum weight masks.
- It is sufficient to calculate bias values only for LCs containing masks with the weight $\varphi(\mathbf{M}^X) = \varphi(\mathbf{M}^Y) \leq n/2$.

To consider the last item in more detail, it is useful to introduce the maximum bias function

$$f_{\max}(t) \overset{def}{=} \max_{\mathbf{M}^X, \mathbf{M}^Y : \varphi(\mathbf{M}^X) = t} b \tag{2.8}$$

for a given weight t. It is easy to see that $f_{\max}(n - t) = f_{\max}(t)$; therefore, it is sufficient to analyze this function only for $t = \{1, \ldots, n/2\}$. For first-order CP boxes with recursive topology $\forall\, \mathbf{M}^X, \mathbf{M}^Y \in \{0, 1\}^n$: $\varphi(\mathbf{M}^X) = \varphi(\mathbf{M}^Y) = 1$, we have $p(\mathbf{M}^X \rightarrow \mathbf{M}^Y) = n^{-1}$; hence $f_{\max}(t) = 1/2n$. For the second-order CP box we have several different types of LCs having biases $b_1 = n^{-2}$, $b_2 = 2n^{-2}$, $b_3 = 3n^{-2}$, and $b_4 = 4n^{-2}$; hence $f_{\max}(t) = 4/n^2$. For the hth order CP box, where $h \leq n$, the $f_{\max}(t)$ function decreases in the interval $1 \leq t \leq n/2$. In the interval $n/2 \leq t \leq n$, the $f_{\max}(t)$ function increases. Let us fix the value $t = t_0$ and consider the $f_{\max}(t)$ values as a function of the order of the recursive CP boxes: $F = F(h) = f_{\max}(t_0, h)$. The function $F(h)$ decreases in the interval $1 \leq h \leq n$ for arbitrary fixed $t \in \{1, 2, \ldots, n/4\}$. Such properties of the DDP operations allow us to make an assumption that for linear attacks against the DDP-based ciphers the most efficient LCs are those that contain masks with a few active bits. Besides, we can expect that the higher the h value, the more the contribution of CP boxes against LCA.

While evaluating the security of several ciphers described in the following text, we will perform LCA from the designer's point of view, i.e., we will try to distinguish the n-bit encryption function from a random transformation using the LCA. If we find an LC the bias value of which is larger than $b_{\mathrm{rand}} = 2^{-n/2}$, we conclude that the cipher is not secure against LCA.

Actually, we should take into account that CP boxes must contribute to the security against other attacks than LCA, for example, against differential analysis, in the which case it is desirable to have significant avalanche effect. The DDP contributes to avalanche when changes are introduced in the controlling data subblock. When a controlling data subblock has a length of n bits, each of its bits influences $s/2$ switching elements of the CP box, where the number of active layers is $s = \log_2 nh$.

2.3 Differential Characteristics

The contribution of CP boxes to the security of the DDP-based ciphers is characterized by differential properties of CP boxes. Let two inputs to the CP box be X' and X'' with corresponding outputs Y' and Y'', where $X', X'', Y', Y'' \in \{0, 1\}^n$. The input difference is $\Delta^X = X' \oplus X''$. It is said that Δ^X transforms into the output difference $\Delta^Y = Y' \oplus Y''$, while passing an operation box. Depending on two input values X' and $X'' = X' \oplus \Delta^X$, the same input difference Δ^X transforms into different output differences Δ^Y with different probabilities $\Pr(\Delta^Y/\Delta^X)$. For random transformation, we have $\Pr(\Delta^Y/\Delta^X) = 2^{-n}$. However, encryption is a deterministic transformation for which $\Pr(\Delta^Y/\Delta^X)$ can significantly exceed the value 2^{-n} for some particular pairs of Δ^X and Δ^Y. The triple $(\Delta^X, \Delta^Y, \Pr(\Delta^Y/\Delta^X))$ is referred to as a *differential characteristic* (DC) of some operation or transformation. The attacker tries to find and exploit the DC with maximum value of the probability $\Pr(\Delta^Y/\Delta^X)$. Such attacks are called differential cryptanalysis (DCA) [5]. An encryption process can be represented as performing several consecutive operations differential properties of which define the characteristics of the total transformation. Thus, the efficiency of the DDP boxes to thwart the differential attack can be estimated considering their differential characteristics. For controlled operations, one should consider the differences $\Delta^{(X,V)} = (\Delta^X, \Delta^V)$ as input ones; DC can be described as $(\Delta^X, \Delta^V, \Delta^Y, \Pr(\Delta^Y/\Delta^X))$.

Nonzero bits of a difference are called *active bits*. Let the lower index in the Δ_t difference denote the number of active bits. In the simplest case $\Delta^V = \Delta_0^V = (0, \ldots, 0)$ for the CP boxes, we have $\Pr(\Delta^Y/\Delta^X) \neq 0$ only if $\varphi(\Delta^Y) = \varphi(\Delta^X)$. The $\Pr(\Delta^Y/\Delta^X)$ value is equal to the probability that input vector Δ^X is transformed into output vector Δ^Y. For CP boxes of the order $h \leq n$ and differences (Δ^Y, Δ^X) satisfying the condition $\varphi(\Delta^Y) = \varphi(\Delta^X) = t \leq h$, we have $\Pr(\Delta^Y/\Delta^X) \leq n^{-1}$. On average, one can estimate $\Pr(\Delta^Y/\Delta^X) = \binom{n}{t}^{-1}$. The highest probability value n^{-1} has differential characteristics with differences containing one active bit: $\varphi(\Delta^Y) = \varphi(\Delta^X) = 1$. To consider the general behavior of the differences in the case $\varphi(\Delta^V) = 0$, it is useful to introduce the maximum probability function

$$f_{\max}(t) \overset{def}{=} \max_{\Delta^X,\Delta^Y:\varphi(\Delta^X)=t} \Pr(\Delta^Y / \Delta^X) \qquad (2.9)$$

for given fixed weight t. If $t > 1$, then the probability distribution of the differences is nonuniform; however, if the order of the CP box increases, then the difference between the maximum and minimum probability decreases, as does the maximum probability value. Therefore, the larger the h value of the CP box, the larger is its contribution to the security against DCA, as in the case of the LCA. As the probability of DC sharply reduces with an increase in the t value, we can expect that the most efficient DCs are those containing differences with few active bits. While performing DCA, the most efficient differences should be selected depending on the concrete structure of the encryption algorithm. At this point, we can discuss only the general tendency that is introduced by the DDP operations.

It is evident that the active bits of input difference Δ^X do not contribute to the avalanche. The avalanche effect connected with the DDP boxes is caused by the active bits of the difference Δ^V_t, where $t \geq 1$. If we have Δ^X_0, then the difference Δ^V_1 (or Δ^V_t, where $t \leq n/8$) generates two (or $2i$, where $i = 1, \ldots, t$) active bits in the Δ^Y output difference with the probability 2^{-1} $\left(\approx \frac{2^{-t} t(t-1)\ldots(t-i+1)}{i!}\right)$. This fact evidently shows that using data subblocks to specify the value V increases significantly the efficiency of CP boxes as cryptographic primitive.

Thus, similarly to the linear attack, the differential cryptanalysis of the DDP-based ciphers seems to be more efficient with the differences, with a few active bits. In the case $\varphi(\Delta^V) = 0$, the probability of the differential characteristic can be calculated as the probability that input active bits are moved to output active bits, as in the case of calculating LC biases.

As the $\mathbf{P}_{2/1}$ box is the main building block in the DDP boxes, the DCs of these boxes are defined by their topology, distribution of the controlling bits, and DCs of the switching element $\mathbf{P}_{2/1}$. Due to the small size of the $\mathbf{P}_{2/1}$ box, it is easy to determine all its differential characteristics (see Figure 2.5). While considering differential properties of the DDP boxes, it is efficient to interpret the process of transforming two input values X' and $X'' = X' \oplus \Delta^X$ as performing the DDP operation on vector Δ^X. With such interpretation, we can characterize the differential properties of the DDP operation, using the probabilities of some sets of DCs with fixed weights $\varphi(\Delta^X) = t_1$ and $\varphi(\Delta^Y) = t_2$. Thus, instead of considering the concrete pairs of input and output differences, sometimes we will consider some free-active-bits DC defined as follows:

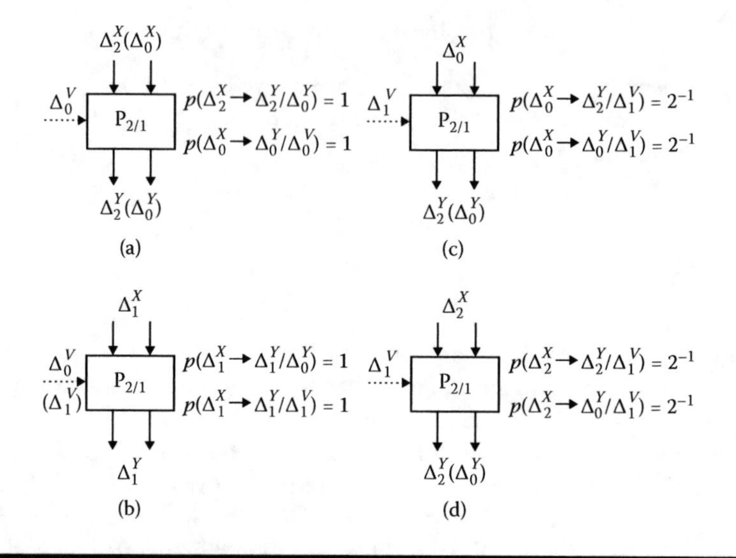

Figure 2.5 Differential characteristics with nonzero probability corresponding to the elementary CP box $\mathbf{P}_{2/1}$.

Definition 2.3. *Suppose U is a binary vector. Free-active-bits difference connected with this vector is an arbitrary difference Δ_t^U such that $\Delta_t^U \in \{\Delta^U : \varphi(\Delta^U) = t\}$.*

To distinguish concrete t-bit differences from free-active-bits differences Δ_t^U having the same weight, the first ones are denoted as $\Delta^U_{t|i_1, i_2, \ldots, i_t}$, where i_1, i_2, ... i_t denote numbers of digits to which the active bits correspond.

Definition 2.4. *Suppose $Y = F(X)$ is a transformation function. Free-active-bit differential characteristic is a triple $\left(\Delta_{t_1}^X, \Delta_{t_2}^Y, \Pr(\Delta_{t_2}^Y / \Delta_{t_1}^X)\right)$, where Δ_{t1}^X and Δ_{t2}^Y are input and output free-active-bits differences, respectively, and $\Pr(\Delta_{t_2}^Y / \Delta_{t_1}^X)$ is the probability of the output difference, provided the input difference is Δ_{t1}^X.*

The notions of free-active-bits DC and free-active-bits differences are useful to characterize some general differential properties of the DDP operations. For example, consideration of the transformation $\Delta_{t_1}^X \xrightarrow{P_{n/m}} \Delta_{t_2}^Y$ corresponds to the transformation of arbitrary input difference $\Delta_{t_1}^X \in \{\Delta^X : \varphi(\Delta^X) = t_1\}$ into arbitrary output difference $\Delta_{t_2}^Y \in \{\Delta^Y : \varphi(\Delta^Y) = t_2\}$, while the $\Delta_{t_1}^X$ difference passes through the $\mathbf{P}_{n/m}$ box.

Note that in the case $\varphi(\Delta^V) = 0$, the free-active-bits DCs $\left(\Delta_t^X, \Delta_t^Y, \Pr(\Delta_t^Y / \Delta_t^X)\right)$ for each possible value t have the probability $\Pr\left(\Delta_t^Y / \Delta_t^X\right) = 1$, because they take into account only the weight of the differences and do not differentiate between different positions of active bits. If $\varphi(\Delta^V) = 0$ and $t_1 \neq t_2$, then $\Pr\left(\Delta^Y_{t_2} / \Delta^X_{t_1}\right) = 0$. The free-active-bits DCs are useful, when the case $\varphi(\Delta^V) \neq 0$ is considered. Using DC of the switching element and taking into account the bit distribution of the controlling data subblock L, it is easy to compute the probabilities of differential characteristics for the most important types of the DDP boxes. Let us consider a case of using extension box **E** that provides for the influence of each bit of L on q ($q = 2, 3$) different output bits of the **E** box. To define the **E** box completely, we should indicate the concrete distribution of bits of the controlling data subblock L. The **E** boxes represent an essential part of the DDP operations. However, at this point, it is sufficient to define some general properties of **E** boxes. We suppose that the following design criteria are satisfied.

Criterion 2.1. *Let* $\mathbf{P}_{n/m}$ *be an s-layer DDP operation transforming an input data block X under the control of the controlling data subblock L. Then, for all values of the controlling data subblock L and* $\forall\, i \in \{1, 2, \ldots, n\}$, *the input bit* x_i *should be permuted depending on s different bits of L.*

Criterion 2.2. *Let a DDP box contain s active layers and* $s \geq 2$ *be an even number. Then each bit of the controlling vector L should control s/2 different switching elements.*

For different concrete cases of the CP box design, one can also take into account some additional requirements of the **E** boxes, for example: (1) mirror symmetry of the distribution of the controlling bits corresponding to the left and right halves of the controlling data subblock and (2) mirror-symmetry position of the switching elements controlled with the same bit of the controlling data subblock.

Suppose that Criterion 2.1 is satisfied. Then each bit of the controlling data subblock influences s different bits of the transformed data subblock. In other words, there exists no input bit that crosses more than one switching element controlled with the same bit of the controlling data subblock. The switching elements controlled with active bits of the difference corresponding to the controlling data are described as *active.* If no active bits of the difference Δ^X cross an active box $\mathbf{P}_{2/1}$, then the last generates two active bits with probability 0.5 (see Figure 2.5*c*). If one active bit of Δ^X crosses the active box $\mathbf{P}_{2/1}$, then this bit passes the switching element with probability 1 (see Figure 2.5*b*). If two active bits of Δ^X are fed to the active box $\mathbf{P}_{2/1}$, then this pair of active bits is annihilated with probability 0.5 (see Figure 2.5*d*). These elementary events and their probabilities define differential properties of the DDP boxes.

Figure 2.6 illustrates the process of active bits' passing through some hypothetical CP box $\mathbf{P}_{n/m}$, for example, $\mathbf{P}_{16/32}$ ($q = 2$), $\mathbf{P}_{32/96}$ ($q = 3$), or $\mathbf{P}_{64/192}$ ($q = 3$). Let the difference Δ^L correspond to the controlling data subblock. Because no transformation of L is performed, the Δ^L difference does not change; however, it influences significantly the Δ^Y output difference of the CP box. The extension box **E** multiplies

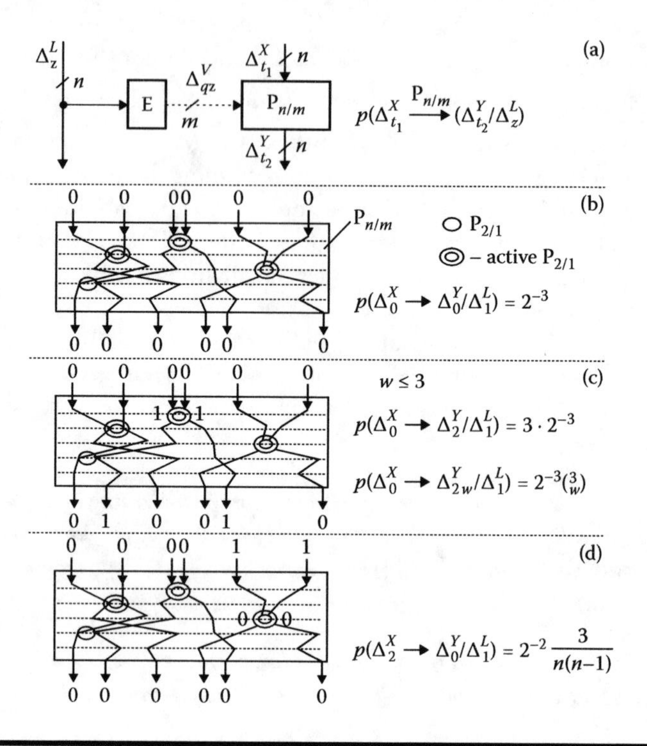

Figure 2.6 A difference $(\Delta^X_{t_1}, \Delta^L_z)$ passing through the CP box: (*a*) general scheme, (*b*) the case $t_1 = t_2 = 0$, (*c*) generation of a pair of active bits, and (*d*) annihilation of a pair of active bits.

each of z active bits of the difference Δ^L producing qz active bits of the controlling vector that activates qz elementary boxes $\mathbf{P}_{2/1}$. Some of these boxes generate pairs of active bits. Suppose $\varphi(\Delta^X) = 0$ and $2qz \ll n$, then the probability that the same bit of X passes two active switching elements is sufficiently low, and we can consider that the events of generation of the active bits pairs are independent. In this case, it is easy to calculate the probability that w pairs of active bits will be generated and get the following formula:

$$p(\Delta_0^X \to \Delta_{2w}^Y / \Delta_z^L) \approx 2^{-qz} \binom{w}{qz} \tag{2.10}$$

If $z = 1$, then this formula is exact.

Another characteristic case relates to the Δ_1^X free-active-bits difference containing one active bit. The active bit of the Δ_1^X difference can cross no active switching elements. Considering that $\Delta_1^X \in \{\Delta^X : \varphi(\Delta^X) = 1\}$, $L \in \{0, 1\}^n$ and $X \in \{0, 1\}^n$, are uniformly distributed random values it is easy to get the following formula for the probability of this event: $p' = (n - 2q)/n$. The probability that active input bit crosses one of the active elements $\mathbf{P}_{2/1}$ is $p'' = 2q/n$. For Δ_z^X difference, where $z = 2, 3, 4$, the p' and p'' probabilities can be estimated approximately using the formulas like those corresponding to the case $z = 1$:

$$p' \approx (n - 2qz)/n \text{ and } p'' \approx 2qz/n \tag{2.11}$$

Taking into account that each active elementary box $\mathbf{P}_{2/1}$ having at its input a zero difference can generate two active bits with a probability of 0.5, it is easy to derive the following approximation for the probability value of the considered event:

$$p(\Delta_1^X \to \Delta_{1+2w}^Y / \Delta_z^L) \approx 2^{-qz}\binom{qz}{w}\frac{(n-2qz)}{n} + 2^{1-qz}\binom{qz-1}{w}\frac{2qz}{n}, \tag{2.12}$$

where $w \leq qz$ and $\binom{qz-1}{qz} \overset{def}{=} 0$

When two or more active bits pass through the CP box, an even number of active bits can be annihilated. The maximum probability corresponds to the case of annihilation of two active bits. Let us consider the differences Δ_2^X and Δ_z^L. With a probability close to $2qz/n$, one of the active bits of the Δ_2^X difference crosses one of the active elementary switches. With a probability of $1/(n - 1)$, the second active bit moves to the input of the same box, $\mathbf{P}_{2/1}$. With probability 0.5, these active bits are simultaneously annihilated. The remaining $qz - 1$ active switching elements generate no pair of active bits with a probability 2^{-qz+1}. Multiplying the probabilities of these three independent events and assuming that $\Delta_2^X \in \{\Delta^X : \varphi(\Delta^X) = 2\}$ is a uniformly distributed random value, we get:

$$p(\Delta_2^X \to \Delta_0^Y / \Delta_z^L) \approx 2^{1-qz}\frac{qz}{n(n-1)}. \tag{2.13}$$

The considered mechanism of annihilating the active bits proves that simultaneous zeroing of two and more pairs of active bits has a considerably smaller probability. Some values of the probability $p(\Delta^X \to \Delta^Y/\Delta_z^L)$ are shown in Tables 2.1 and 2.2.

2.4 Cobra-H64: A 64-Bit Block Cipher Based on Variable Permutations

2.4.1 Specification of the Encryption Algorithm

A study by Goots et al. [14] has introduced the 64-bit block cipher SPECTR-H64 designed by combining a Feistel's network with permutation networks. The

Table 2.1 Probabilities $p(\Delta^X \rightarrow \Delta^Y/\Delta_1^L)$ for the second-order box $\mathbf{P}_{64/192}$

	Δ_0^X	Δ_1^X	Δ_2^X
Δ_0^Y	2^{-3}	0	$1.52 \cdot 2^{-13}$
Δ_1^Y	0	$1.1 \cdot 2^{-3}$	0
Δ_2^Y	$1.5 \cdot 2^{-2}$	0	$1.11 \cdot 2^{-3}$
Δ_3^Y	0	$1.55 \cdot 2^{-2}$	0
Δ_4^Y	$1.5 \cdot 2^{-2}$	0	$1.42 \cdot 2^{-2}$
Δ_5^Y	0	$1.45 \cdot 2^{-2}$	0

Table 2.2 Probabilities $p(\Delta^X \rightarrow \Delta^Y/\Delta_2^L)$ for the second-order box $\mathbf{P}_{64/192}$

	Δ_0^X	Δ_1^X	Δ_2^X
Δ_0^Y	2^{-6}	0	$1.52 \cdot 2^{-15}$
Δ_1^Y	0	$1.19 \cdot 2^{-6}$	0
Δ_2^Y	$1.5 \cdot 2^{-4}$	0	$1.1 \cdot 2^{-6}$
Δ_3^Y	0	$1.69 \cdot 2^{-4}$	
Δ_4^Y	$1.88 \cdot 2^{-3}$	0	$1.51 \cdot 2^{-4}$
Δ_5^Y	0	2^{-2}	0
Δ_6^Y	$1.25 \cdot 2^{-2}$	0	$1.72 \cdot 2^{-3}$
Δ_7^Y	0	$1.25 \cdot 2^{-2}$	0

peculiarity of the round transformation of SPECTR-H64 is the use of two mutually inverse DDP operations of the first order: $\mathbf{P}_{32/80}$ and $\mathbf{P}^{-1}_{32/80}$. Because the controlling data subblock is not transformed in the round transformation, SPECTR-H64 requires modification of the controlling vectors of the $\mathbf{P}_{32/80}$ and $\mathbf{P}^{-1}_{32/80}$ boxes with subkeys of a different round. Besides, the $\mathbf{P}_{32/80}$ and $\mathbf{P}^{-1}_{32/80}$ boxes contain five active layers; therefore, the controlling data subblock L has an unbalanced influence on the formation of the controlling vectors corresponding to the indicated CP boxes. To improve that design, the Cobra-H64 block cipher has been proposed [42, 65]. Cobra-H64 uses the second-order boxes $\mathbf{P}_{32/96}$ and $\mathbf{P}^{-1}_{32/96}$ containing six active layers, making it possible to balance the controlling data subblock's influence on DDP permutations. In one round of Cobra-H64, the subblock L is transformed; therefore, no keys are needed to be mixed with L. The structure of the used $\mathbf{P}_{32/96}$ and

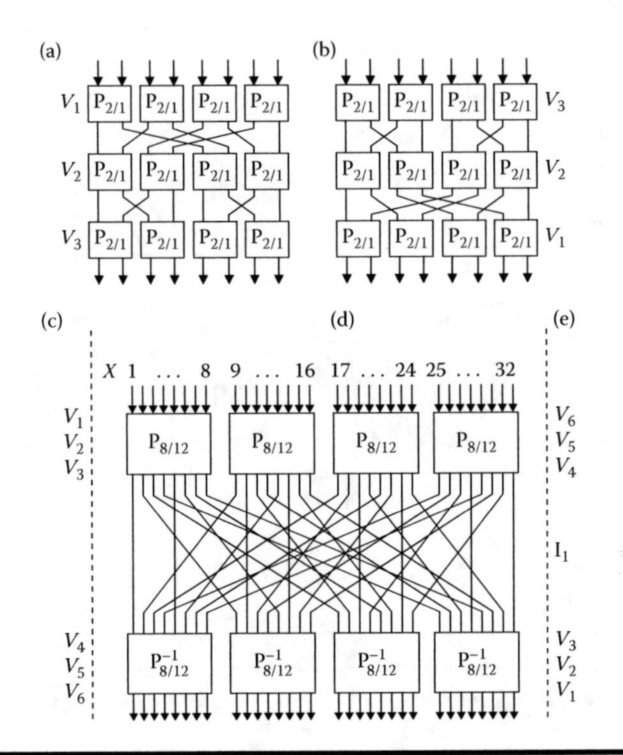

Figure 2.7 Structure of the CP boxes $P_{8/12}$ (*a*), $P^{-1}_{8/12}$ (*b*), $P_{32/96}$ (*c, d*), and $P^{-1}_{32/96}$ (*e, d*).

$\mathbf{P}^{-1}_{32/96}$ boxes is presented in Figure 2.7. They have symmetric topology that differs from the recursive topology of the second-order box $\mathbf{P}_{32/96}$. Due to the mirror symmetry of their topology, the $\mathbf{P}_{32/96}$ and $\mathbf{P}^{-1}_{32/96}$ boxes of the Cobra-H64 cipher differ from each other only in the distribution of controlling bits. They are constructed using a cascade containing four $\mathbf{P}_{8/12}$ boxes and a cascade containing four $\mathbf{P}^{-1}_{8/12}$ boxes (see Figure 2.7*a* and 2.7*b*). The output of the upper cascade is connected with the input of the lower cascade, by means of the permutational involution $\mathbf{I}_1$ described as follows:

(1)(2,9)(3,17)(4,25)(5)(6,13)(7,21)(8,29)(10)(11,18)
(12,26)(14)(15,22)(16,30)(19)(20,27)(23)(24,31)(28)(32).

While designing the single-key cryptosystem Cobra-H64, the design criteria used were the following:

1. The cryptosystem should be an iterated 64-bit cipher.
2. The cryptalgorithm should be able to perform encryption and decryption with a simple and fast change of the sequence of used subkeys.

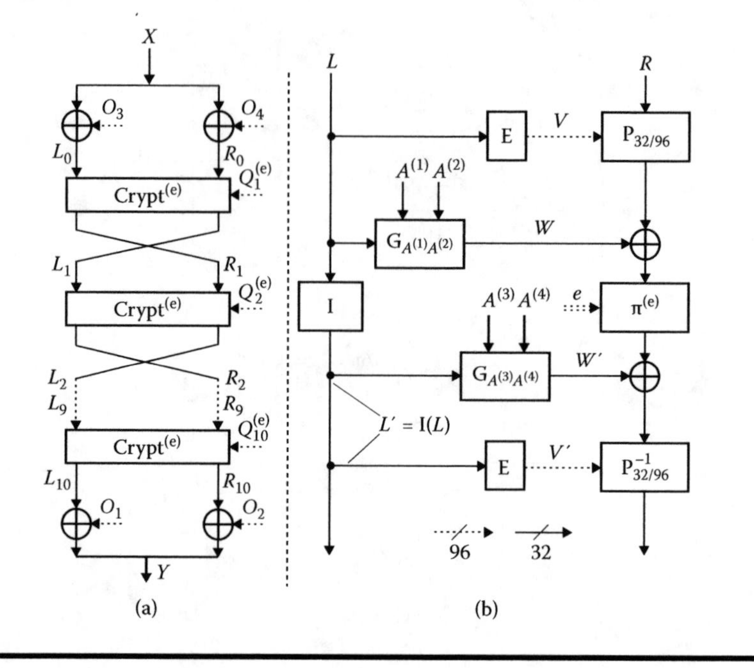

Figure 2.8 General structure of Cobra-H64 (*a*) and procedure Crypt$^{(e)}$ (*b*).

3. The cipher should be fast in case of frequent change of keys. For this reason, a very simple key scheduling is used.
4. Round transformation of data subblocks should be characterized by high parallelism.

Let us denote the input data block as $X = (x_1, x_2, \ldots, x_{64}) = (L, R)$, where $L = (x_1, x_2, \ldots, x_{32})$ and $R = (x_{33}, x_{34}, \ldots, x_{64})$. Let $e = 0$ denote encryption and $e = 1$ denote decryption. Cobra-H64 is a new 10-round iterated block cipher with 64-bit input and 128-bit secret key. The general iterative structure of Cobra-H64 and the round transformation are presented in Figure 2.8. Encryption and decryption are described by the following general equations:

$$Y = \mathbf{T}^{(e=0)}(X, K) \text{ and } X = \mathbf{T}^{(e=1)}(Y, K), \tag{2.14}$$

where Y is the ciphertext and K is the secret key represented as a concatenation of four subkeys , for $K_i \in \{0, 1\}^{32}$: $K = (K_1, K_2, K_3, K_4)$. Cobra-H64 uses no preprocessing to transform the subkeys. The extended key $Q^{(e)} = (Q_1^{(e)}, Q_2^{(e)}, \ldots, Q_{10}^{(e)})$, where $Q_j^{(e)}, j = 1, 2, \ldots, 10$, are round keys, is formed as a simple sequence of subkeys K_i taken in the order specified by the key scheduling table. Encryption begins with initial transformation. Then 10 rounds with **Crypt**$^{(e)}$ procedure followed by the final transformation are performed. The initial transformation consists in XORing

Table 2.3 Specification of the round subkeys $A^{(i)}$ in Cobra-H64

$j =$	1	2	3	4	5	6	7	8	9	10
$A_j^{(1)} =$	Q_1	Q_4	Q_3	Q_2	Q_1	Q_1	Q_2	Q_3	Q_4	Q_1
$A_j^{(2)} =$	Q_2	Q_1	Q_4	Q_3	Q_4	Q_4	Q_3	Q_4	Q_1	Q_2
$A_j^{(3)} =$	Q_3	Q_2	Q_1	Q_4	Q_3	Q_3	Q_4	Q_1	Q_2	Q_3
$A_j^{(4)} =$	Q_4	Q_3	Q_2	Q_1	Q_2	Q_2	Q_1	Q_2	Q_3	Q_4

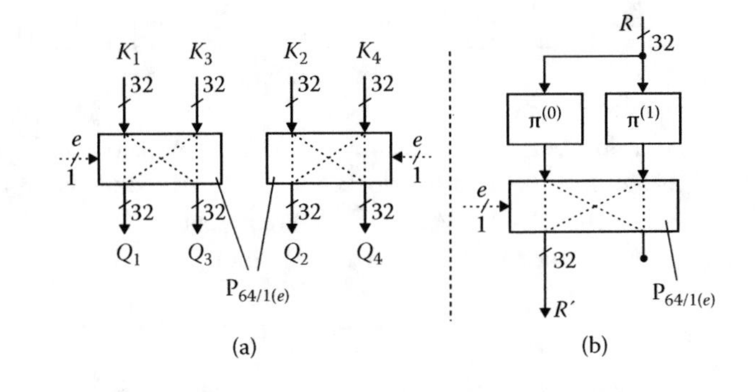

Figure 2.9 Swapping operations of Cobra-H64.

subkeys with data subblocks: $L_0 = L \oplus O_3$ and $R_0 = R \oplus O_4$. The final transformation is $L_{11} = L_{10} \oplus O_1$ and $R_{11} = R_{10} \oplus O_2$. The ciphertext block is $Y = (L_{11}, R_{11})$.

Each of the round keys $Q_j^{(e)}$ consists of four e-dependent round subkeys $A^{(1)}$, $A^{(2)}, A^{(3)}, A^{(4)} \in \{0, 1\}^{32}$, i.e., $Q_j^{(e)} = (A^{(1)}, A^{(2)}, A^{(3)}, A^{(4)})_j^{(e)}$. Table 2.3 and Figure 2.9*a* specify the round subkeys and their correspondence to the secret key. The e-dependent round subkeys Q_i ($i = 1, 2, 3, 4$) are generated at the outputs of two transposition boxes each of which is a single-layer box $\mathbf{P}_{64/1}^{(e)}$. In the box $\mathbf{P}_{64/1}^{(e)}$, all elementary switching elements are controlled with the same bit e. The pairs (K_1, K_3) and (K_2, K_4) are fed to the corresponding boxes $\mathbf{P}_{64/1}^{(e)}$. For $e = 0$, we have $Q_i = K_i$. For $e = 1$, we have $Q_1 = K_3$, $O_2 = K_4$, $O_3 = K_1$, $O_4 = K_2$.

The e-dependent fixed permutation $\pi^{(e)}$ in the procedure $\mathbf{Crypt}^{(e)}$ (see Figure 2.9*b*) is also important to provide correct decryption process. This switchable permutation $\pi^{(e)}$ is implemented transposing the outputs of two fixed permutations $\pi^{(0)}$ and $\pi^{(1)}$ with a single-layer box $\mathbf{P}_{64/1}^{(e)}$. Permutations $\pi^{(0)}$ and $\pi^{(1)}$ contain two cycles. The first cycle corresponds to the identical permutation of the input bit r_{32}. The second cycle is described as follows:

$$\pi^{(0)}(r_1, r_2, ..., r_{31}) = (r_1, r_2, ..., r_{31})^{<<<5}, \tag{2.15}$$

$$\pi^{(1)}(r_1, r_2, ..., r_{31}) = (r_1, r_2, ..., r_{31})^{<<<26}. \tag{2.16}$$

Table 2.4 Distribution of the controlling data subblock bits in the $\mathbf{P}_{32/96}$ box (indices of the controlling bits)

V_1	1	2	3	4	5	6	7	8	9	10	11	12	13	14	15	16
V_2	7	8	9	10	11	12	13	14	15	16	1	2	3	4	5	6
V_3	13	14	15	16	1	2	3	4	5	6	7	8	9	10	11	12
V_4	17	18	19	20	21	22	23	24	25	26	27	28	29	30	31	32
V_5	23	24	25	26	27	28	29	30	31	32	17	18	19	20	21	22
V_6	29	30	31	32	17	18	19	20	21	22	23	24	25	26	27	28

Controlling vectors corresponding to CP boxes $\mathbf{P}^{(V)}_{32/96}$ and $(\mathbf{P}^{-1}_{32/96})^{(V')}$ are formed using the same extension box **E** implemented as simple connections. Suppose the current value of the controlling data subblock $L = (L_l, L_h)$ is fed to the input of the **E**-box. Then, its output $V = (V_1, V_2, V_3, V_4, V_5, V_6)$ represented as a concatenation of six 16-bit vectors V_i is specified by the following equations:

$$V_1 = L_l,\ V_2 = L_l^{<<<6},\ V_3 = L_l^{<<<12},\ V_4 = L_h,\ V_5 = L_h^{<<<6},\ V_6 = L_h^{<<<12} \quad (2.17)$$

Such a distribution of the controlling vectors satisfies Criteria 2.1 and 2.2. Actually, for arbitrary given vector L, the permutation of each input bit of the CP box is defined by six different bits of L, and each bit of L defines three different bits of V. Thus, in one round, each bit of L influences three switching elements in the box $\mathbf{P}^{(V)}_{32/96}$ and three switching elements in the box $(\mathbf{P}^{-1}_{32/96})^{(V')}$. The arbitrary input bit of the boxes $\mathbf{P}_{32/96}$ and $\mathbf{P}^{-1}_{32/96}$ moves to each output position with the same probability provided L is a uniformly distributed random variable. Table 2.4 describes the distribution of the controlling bits in evident form.

In Cobra-H64, the element of the Feistel-like cipher is represented by two **G** boxes each of which represents a controlled two-place operation (CTPO). To gain better spreading of the avalanche effect caused by the **G** operations, two fixed permutations (**I** and $\pi^{(e)}$) are used. The $\pi^{(e)}$ operation is described in the previous text. The **I** operation performed on the left data subblock L is the following permutational involution:

(1,17)(2,21)(3,25)(4,29)(5,18)(6,22)(7,26)(8,30)(9,19)
(10,23)(11,27)(12,31)(13,20)(14,24)(15,28)(16,32).

This notation of the permutation evidently shows its cyclic structure. We see 16 cycles of the length 2. In general, we have cycles of different length, for example (33, 37, 41, 39, 33). Each number in every cycle represents the bit that is permuted to the digit indicated by the next number (i.e., its right neighbor). The rightmost number stands for the bit that is permuted to the digit indicated by the leftmost number.

Table 2.5 Formulas describing the avalanche in the G operation corresponding to changing one input bit l_i ($\Delta l_i = 1$)

#	Expression	Probability
1	$\Delta w_i = \Delta l_i$	$p(\Delta w_i = 1) = 1$
2	$\Delta w_{i+1} = \Delta l_i (l_{i-1} \oplus l_{i-2} \oplus a''_{i+1} l_{i-1})$	$p(\Delta w_{i+1} = 1) = 0.5$
3	$\Delta w_{i+2} = \Delta l_i (l_{i-1} \oplus l_{i+1} \oplus a''_{i+1} \oplus a''_{i+2} l_{i+1})$	$p(\Delta w_{i+2} = 1) = 0.5$
4	$\Delta w_{i+3} = \Delta l_i (l_{i+1} \oplus l_{i+2} \oplus a'_{i+2})$	$p(\Delta w_{i+3} = 1) = 0.5$

Note: The probabilities indicated relate to the uniformly distributed random variable L.

The **I** permutation satisfies the following conditions. Let $y_{i'}$ and $y_{j'}$ be the output bits corresponding to the input bits x_i and x_j. Then:

1. $\forall\, i, j: |j - i| \le 3$ we have $|j' - i'| \ge 4$.
2. $\forall\, i$ we have $|i - i'| \ge 6$.

These conditions are used to define that each bit of the left subblock influences as many bits as possible after the outputs of both **G** operations are XORed with R. The **I** involution provides that changing one bit of the left data subblock causes inversion of bits 2 to 8, of the right data subblock, after the outputs of both **G** operations are XORed with R. The permutation $\pi^{(e)}$ introduces no violation to this rule, because $\pi^{(e)}$ shifts the bits of R by 5 digits—except the rightmost one, which is not shifted. After combining the outputs of the **G** operations with the right data subblock, 2 bits of R change deterministically and 6 bits change statistically, with probability 1/2 (see Table 2.5).

Nonlinear G operation. The output of the **G** operation is described by the following expression:

$$W = L_0 \oplus A'_0 \oplus L_2 \otimes L_3 \oplus L_1 \otimes L_2 \oplus L_1 \otimes L_3 \oplus L_2 \otimes A''_1 \oplus A'_1 \otimes L_3 \oplus A''_0 \otimes L_1 \otimes L_2 \quad (2.18)$$

where binary vectors L_j, A'_j, and A''_j are expressed as follows:

$$L_0 = L = (l_1, l_2, \ldots, l_{32}),\ L_1 = (1, l_1, l_2, \ldots, l_{31}) \quad (2.19)$$

$$L_2 = (1, 1, l_1, l_2, \ldots, l_{30}),\ L_3 = (1, 1, 1, l_1, l_2, \ldots, l_{29}) \quad (2.20)$$

$$A'_0 = A' = (a'_1, a'_2, \ldots, a'_{32}),\ A'_1 = (1, a'_1, a'_2, \ldots, a'_{31}) \quad (2.21)$$

$$A''_0 = A'' = (a''_1, a''_2, \ldots, a''_{32}),\ A''_1 = (1, a''_1, a''_2, \ldots, a''_{31}) \quad (2.22)$$

$$A''_2 = (1, 1, a''_1, a''_2, \ldots, a''_{30}) \quad (2.23)$$

For the upper (lower) **G** operation, we have $A' = A^{(1)}$ and $A'' = A^{(2)}$ ($A' = A^{(3)}$ and $A'' = A^{(4)}$). The *i*th bit of the output binary vector W is the following Boolean function:

$$w_i = l_i \oplus a'_i \oplus l_{i-2}l_{i-3} \oplus l_{i-1}l_{i-2} \oplus l_{i-1}l_{i-3} \oplus l_{i-2}a''_{i-1} \oplus a'_{i-1}l_{i-3} \oplus a''_i l_{i-1}l_{i-2} \quad (2.24)$$

where l, a'_i and a''_i are the components of the vectors L, A', and A'', respectively, $(l_{-1}, l_{-2}, l_{-3}) = (1, 1, 1)$, and $a'_0 = a''_0 = 1$.

Peculiarities of Cobra-H64. In general, cipher Cobra-H64 presents an example of extensive use of the CP-box operations. They are used in three different ways:

- As DDPs that are one of two basic cryptographic primitives.
- To swap subkeys when changing from the encryption to the decryption mode.
- To switch permutation $\pi^{(e)}$ when changing the ciphering mode.

The second basic cryptographic primitive is represented by the **G** operation. Due to the high parallelism of the general structure of Cobra-H64, it takes only about $15t_\oplus$ to perform one round of Cobra-H64, where $t_\oplus$ is the delay time of the XOR operation. Time delay of ten rounds is about 150 $t_\oplus$. Encryption speed can be estimated as ≈0.43 bits/$t_\oplus$. The cryptosystem Cobra-H64 is fast in case of frequent change of keys, as it is free of key preprocessing. Cobra-H64 has the following features:

1. In each round it uses the secret key entirely.
2. Internal key scheduling in Cobra-H64 is implemented with two CTPOs executed in parallel with the CP-box operation $\mathbf{P}_{32/96}$.
3. Round transformation includes the special permutational involution performed on the left data subblock.
4. Round transformation includes the special switchable permutation $\pi(e)$ performed on the right data subblock. Permutation $\pi(e)$ prevents the appearance of weak keys with the structure $K = (X, X, X, X)$, where $X \in \{0, 1\}^{32}$.
5. After the *i*th round, each bit of the R_{i-1} data subblock influences statistically all bits of L_i.
6. For operations **G**, $\mathbf{P}_{32/96}$, and $\mathbf{P}^{-1}_{32/96}$, it is sufficiently easy to calculate the characteristics corresponding to the differences with a few active bits.

2.4.1 Security Estimation

Differential analysis seems to be the most efficient attack against Cobra-H64. To evaluate security against DCA, we will use the DCs of CP boxes presented in Section 2.2 and avalanche properties of the **G** operation. The avalanche effect corresponding to DDP is caused by the use of data to define the values of V and V'.

Table 2.6 Probabilities $\Pr\left(\Delta^L_{1|i} \xrightarrow{G} \Delta^W_{2|i_1,i_2}\right)$ and $\Pr\left(\Delta^L_{1|i} \xrightarrow{G} \Delta^W_s\right)$ (number of output bits depending on the *i*th input bit)

$i =$	1 to 29	30	31	32
$u =$	4	3	2	1
$\Delta^L_{1\|i} \to \Delta^W_{1\|i}$	2^{-3}	2^{-2}	2^{-1}	1
$\Delta^L_{1\|i} \to \Delta^W_{2\|i,i+1}$	2^{-3}	2^{-2}	2^{-1}	—
$\Delta^L_{1\|i} \to \Delta^W_{2\|i,i+2}$	2^{-3}	2^{-2}	—	—
$\Delta^L_{1\|i} \to \Delta^W_{2\|i,i+3}$	2^{-3}	—	—	—
$\Delta^L_{1\|i} \to \Delta^W_{2}$	1.52^{-2}	2^{-1}	2^{-1}	—
$\Delta^L_{1\|i} \to \Delta^W_{3}$	1.52^{-3}	2^{-2}	—	—
$\Delta^L_{1\|i} \to \Delta^W_{4}$	2^{-3}	—	—	—

Each bit of the left data subblock influences 3 bits of each controlling vector. Each controlling bit influences 2 bits of the right data subblock. Thus, due to DDP, 1 bit of L influences statistically about 12 bits of R.

The avalanche effect corresponding to the **G** operation is defined by its structure, which provides that each bit of the set $\{l_1, l_2, ..., l_{29}\}$ influences four output bits. Bits from l_{30} to l_{32} influence different numbers (u) of output bits. When performing the **G** operation, the change of the ith input bit causes a deterministic change of the ith output bit and a probabilistic change of the $(i + 1)$th, $(i + 2)$th, and $(i + 3)$th output bits. Using the formulas presented in Table 2.5 we can calculate the probabilities $\Pr\left(\Delta^L_{1|i} \xrightarrow{G} \Delta^W_{s|i_1,i_2,...,i_s}\right)$ of the DC containing input differences with one active bit. The results are presented in Table 2.6.

The best known DCA against Cobra-H64 uses the differences with a few active bits [42]. That attack corresponds to a two-round iterative DC with the input difference $\Delta^X = \left(\Delta^L_0, \Delta^R_{1|k}\right)$, where $1 \le k \le 32$. For all k, the difference $\left(\Delta^L_0, \Delta^R_{1|k}\right)$ passing through two rounds transforms with probability $P(2)$ into the output difference $\Delta^Y = \left(\Delta^L_0, \Delta^R_{1|j}\right)$, where $1 \le j \le 32$. In the first round, the active bit passes the right branch with probability 1. In the second round, the active bit passes the left branch and does not generate any active bits in the right branch with probability $P(2)$. The following two cases contribute mostly to the probability of this two-round characteristic.

Case 1. Each of the two **G** operations produces at its output the difference with one active bit. The $\mathbf{P}_{32/96}$ box produces the transformation $\Delta^R_0 \to \Delta^R_{2|i,j}$, where

$j' = \pi^{(e\oplus 1)}(j)$ and $j = \mathbf{I}(i)$. Then, the difference $\Delta^{R}_{2|i,j}$ after XORing with two output differences of both the **G** operations gives a zero value. The zero difference passes through the $\mathbf{P}^{-1}_{32/96}$ box with probability 2^{-3}.

Case 2. Each of the two **G** operations produces at its output the difference with one active bit. The $\mathbf{P}_{32/96}$ box produces at the output zero difference. Then XORing two outputs of the **G** operation with the right data subblock produces a two-bit difference $\Delta^{R}_{2|i,j}$, where $i = \pi^{(e)}(i)$ and $j = \mathbf{I}(i)$. Then, passing through the $\mathbf{P}^{-1}_{32/96}$ box, $\Delta^{R}_{2|i,j}$ transforms into zero difference. Each of the both cases ends with swapping data subblocks.

Let us denote the probability of the event $\left(\Delta^{L}_{1},\Delta^{R}_{0}\right)\rightarrow\left(\Delta^{L}_{0},\Delta^{R}_{1}\right)$ corresponding to Cases 1 and 2 as P' and P'', respectively, i.e., $P(2) \approx P' + P''$. (We neglect the small contribution of other events corresponding to the formation of two and more pairs of active bits at outputs of some active $\mathbf{P}_{2/1}$ boxes in one of the CP boxes $\mathbf{P}_{32/96}$ or $\mathbf{P}^{-1}_{32/96}$.)

Let us consider Case 1, shown in Figure 2.10. The difference passes through two decryption rounds in the following way. In the first round, the difference $\Delta^{R}_{1|k}$ goes through the right branch of the cryptoscheme and transforms into the difference $\Delta^{R}_{1|i}$ ($i \in \{1, ..., 32\}$) with probability 1. After swapping the subblocks, we have $\Delta^{L}_{1|i}$ and Δ^{R}_{0}. In the second round, the active bit goes through the left branch producing the difference $\Delta^{W}_{1|i}$ at the output of the upper **G** operation (event A_1) with probability $p_1 = 2^{-3}$ and difference $\Delta^{W'}_{1|j}$ at the output of the lower **G** operation (event A_2) with probability p_2. At the same time, the difference $\Delta^{L}_{1|i}$ causes the formation of the two-bit difference $\Delta^{R}_{2|i,j'}$ at the output of the operation $\mathbf{P}_{32/96}$ (event A_3) with probability p_3. After being XORed with difference $\Delta^{W}_{1|i}$, the difference $\Delta^{R}_{2|i,j}$ transforms in $\Delta^{R}_{1|j'}$. Passing fixed permutation $\pi^{(1)}$, the difference $\Delta^{R}_{1|j'}$ transforms deterministically in $\Delta^{R}_{1|j}$ producing zero difference Δ^{R}_{0} after XORing with the output of the lower **G** operation. The zero difference passes through the operation $\mathbf{P}^{-1}_{32/96}$ (event A_4) with probability $p_1 = 2^{-3}$. After swapping, we have the difference $\left(\Delta^{L}_{0},\Delta^{R}_{1|j}\right)$ at the output of the second round.

The values p_1, p_2, p_3, and p_4 are presented in Table 2.7, in which the rows correspond to different subsets of numbers i, and p_i is the probability that for a given $\Delta^{L}_{1|i}$ the zero difference passes through the right branch of the second round. Each subset is characterized by equal values p_1, p_2, p_3, and p_4. The probability

$$p_i = \Pr\left((\Delta^{L}_{1|i},\Delta^{R}_{0}) \xrightarrow{\text{round 2}} (\Delta^{L}_{0},\Delta^{R}_{1|j})\right) = p_1 p_2 p_3 p_4 \qquad (2.25)$$

corresponds to the event that for a given i after the second round, swapping left and right data subblocks, the input difference $\left(\Delta^{L}_{1|i},\Delta^{R}_{0}\right)$ transforms into the output

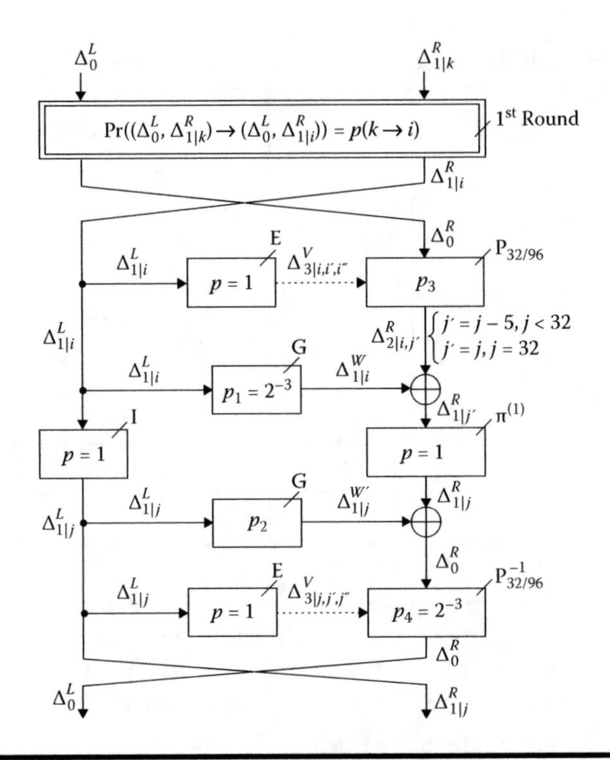

Figure 2.10 Difference $\left(\Delta_0^L, \Delta_{1|k}^R\right)$ passing through two decryption rounds (Case 1).

difference $\left(\Delta_0^L, \Delta_{1|j}^R\right)$, where i is fixed and $j = \mathbf{I}(i)$. Considering that the key and the left data subblock are independent random values, it is easy to see that events A_1, A_2, A_3, and A_4 are independent, pairwise. In the estimation given as follows, we neglect some weak interdependence between the events A_3 and A_4. The probability P' can be calculated with the following formula:

$$P' = p(k \to i) \sum_{i=1}^{32} p_i \approx 1.33 \cdot 2^{-20} \tag{2.26}$$

For $i = 13$, Table 2.7 gives "approximated" values p_1, p_2, and p_3. (The calculation of the contribution of the value $i = 13$ has some peculiarities, because the output bits w_{15} and w_{16} of the upper **G** operation after permutation $\pi^{(1)}$ are XORed with the output bits w'_{20} and w'_{21} of the lower **G** operation. Thus, for $i = 13$, the differences $\Delta_{2|13,15}^R$, $\Delta_{2|13,16}^R$, $\Delta_{2|13,17}^R$, and $\Delta_{2|13,18}^R$ at the output of the $\mathbf{P}_{32/96}$ box contribute to P'. The resultant contribution is $\approx 1.7 \cdot 2^{-23}$.)

The calculation of the probability p_3 is not difficult, but one needs to consider each elementary switching element of the $\mathbf{P}_{32/96}$ box and all connections between

Table 2.7 Probabilities corresponding to different values $i \in \{1, ..., 32\}$

i	p_1	p_2	p_3	p_4	p_i	$p(k \rightarrow i)$
1; 5	2^{-3}	2^{-3}	$1.06{\cdot}2^{-9}$	2^{-3}	1.06^{-18}	2^{-5}
2; 9	2^{-3}	2^{-3}	2^{-13}	2^{-3}	2^{-22}	2^{-5}
8	2^{-3}	2^{-2}	$1.25{\cdot}2^{-11}$	2^{-3}	$1.25{\cdot}2^{-19}$	2^{-5}
12	2^{-3}	2^{-1}	$1.25{\cdot}2^{-11}$	2^{-3}	$1.25{\cdot}2^{-18}$	2^{-5}
13	2^{-2}	2^{-2}	$1.7{\cdot}2^{-11}$	2^{-3}	$1.7{\cdot}2^{-18}$	2^{-5}
16	2^{-3}	1	$1.25{\cdot}2^{-11}$	2^{-3}	$1.25{\cdot}2^{-17}$	2^{-5}
17 to 32	$\neq 0$	2^{-3}	0	2^{-3}	0	2^{-5}
other i	2^{-3}	2^{-3}	$1.25{\cdot}2^{-11}$	2^{-3}	$1.25{\cdot}2^{-20}$	2^{-5}

active layers of the last. For example, let us consider the difference $\Delta^{L}_{1|5}$. The 5th bit controls one elementary box $\mathbf{P}_{2/1}$ in each of the three upper active layers of the CP box $\mathbf{P}_{32/96}$ (see Table 2.4), namely the 5th, 31st, and 41st $\mathbf{P}_{2/1}$ boxes. For $i = 5$, we have $j = 18$ in correspondence with the operations $\mathbf{I}$ and $\pi^{(1)}$. The following three different variants corresponding to generation of the output difference $\Delta^{R}_{2|5,13}$ are to be taken into account:

1. Depending on the value of R, the input bits of the 5th elementary box are different, with probability $p = 0.5$, whereas the input bits corresponding to both the 31st and 41st elementary boxes are equal ($p = 0.5$).
2. The input bits corresponding to the 31st elementary box are different ($p = 0.5$), whereas the input bits corresponding to both the 5th and 41st elementary boxes are equal ($p = 0.5$).
3. The input bits corresponding to the 41st elementary box are different ($p = 0.5$), whereas the input bits corresponding to both the 5th and 31st elementary boxes are equal ($p = 0.5$).

Because each bit of the subblock L controls six different bits of R, the probability of each of these events is exactly 2^{-3}. One of the output bits of the fifth elementary box moves to the fifth digit at the output of CP box $\mathbf{P}_{32/96}$, with probability 2^{-5} (it must pass five active layers having the single route). The same probability corresponds to the second output bit of the 5th elementary box to be moved to the 13th output of $\mathbf{P}_{32/96}$. Thus, we have the probability $p' = (2^{-5})^2 \cdot 2^{-3} = 2^{-13}$ that the first event generates the difference $\Delta^{R}_{2|5,13}$. The probability that the second event causes the difference $\Delta^{R}_{2|5,13}$ is $p'' = 0$ (the input bits of the 31st elementary box never move simultaneously to the 5th and 13th output digits of $\mathbf{P}_{32/96}$). The probability that the third event produces the difference $\Delta^{R}_{2|5,13}$ is $p''' = 2^{-9}$. The resultant probability that the difference $\Delta^L_{1|5}$ leads to $\Delta^R_{2|5,13}$ is $p \approx p' + p'' + p''' \approx 1.06 \cdot 2^{-9}$. Similarly, one can consider the difference $\Delta^L_{1|i}$ for $i = 1, \ldots, 32$. Note that the difference $\Delta^L_{1|i}$, where $17 \leq i \leq 32$, does not contribute to P'.

Case 2 is similar to Case 1, except that the upper CP box produces zero output difference and two active bits are annihilated with probability 0.5 when they pass through the same active switching element $\mathbf{P}_{2/1}$ of the lower CP box. The contribution of this case can be calculated as in Case 1. For Case 2, one can obtain $P'' \approx 2^{-20}$ and then $P(2) \approx P' + P'' \approx 1.16 \cdot 2^{-19}$. This analysis applied to SPECTR-H64 [8,13] has given the following probability for the two-round iterative DC: $P(2) \approx 1.15 \cdot 2^{-13}$. Using the values $P(2)$ we find that Cobra-H64 with $r \geq 8$ encryption rounds and SPECTR-H64 with $r \geq 10$ rounds are indistinguishable from a random cipher with differential attack. A linear attack is less efficient than DCA (on a linear attack against SPECTR-H64 see Ko et al. [22]). To thwart LCA, six rounds of Cobra-H64 are sufficient.

Some comments on the key scheduling should be given. The key scheduling used is secure against basic related-key attacks. In spite of simplicity of the key schedule, the keys $K' = (X, Y, X, Y)$ or $K'' = (X, X, X, X)$, where $X, Y \in \{0, 1\}^{32}$, are not weak, because encryption and decryption algorithms are different (they use different operations $\pi^{(0)}$ and $\pi^{(1)}$). The application of the switchable permutation $\pi^{(e)}$ in Cobra-H64 is only a particular example showing the role of switchable operations in iterative ciphers with simple key scheduling. In general, such operations are very effective in preventing weak or semiweak keys, while simple key scheduling is used.

2.5 DDP-64: Pure DDP-Based Cipher

Variable permutations used to perform the reversible transformations are linear cryptographic primitives; therefore, they should be combined with some nonlinear operations, while being used in designing encryption algorithms. In this approach, it is difficult to insist that variable permutations are the basic primitive contributing mostly to the security of the designed cipher. To demonstrate the efficiency of DDPs in a pure form, they should be combined only with fixed permutations and the XOR operation. In other words, the design of some pure DDP-based ciphers that are secure against known attacks represents a demonstration of the efficiency of DDPs as cryptographic primitives. A variant of such ciphers, called DDP-64, has been designed and investigated in the literature [47,49]. The design strategy of the single-key cryptosystem DDP-64 was oriented to the extensive use of variable permutations that are fast, and inexpensive in hardware implementation. The main feature of the DDP-64 design is the implementation of the nonlinear DDP operations; i.e., DDP-64 uses two types of variable permutations: linear and nonlinear.

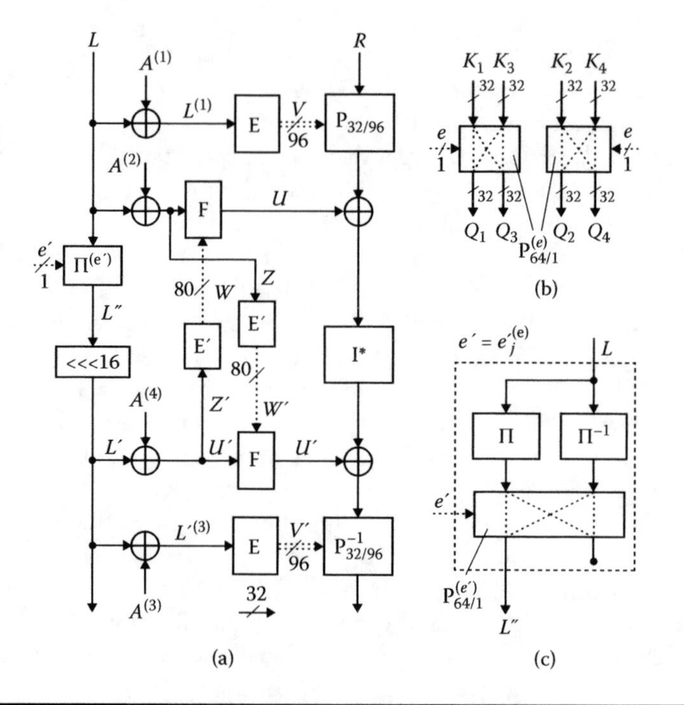

Figure 2.11 Round transformation of DDP-64 (*a*), swapping subkeys (*b*), and structure of the switchable fixed permutation (*c*).

2.5.1 Description of the Encryption Algorithm

DDP-64 is a ten-round iterated block cipher, operating on 64-bit data blocks and using a 128-bit key. The general encryption scheme of DDP-64 is the same as that of Cobra-H64 (see Figure 2.8*a*)— except for the key scheduling. The encryption and decryption functions of DDP-64 can be described by the following equations:

$$Y = F^{(e=0)}(X, K) \text{ and } X = F^{(e=1)}(Y, K) \tag{2.27}$$

where X is the input data block (it is plaintext for encryption), Y is the output data block (it is ciphertext for decryption), $X, Y \in \{0, 1\}^{64}$, K is the secret key ($K \in \{0, 1\}^{128}$), F is the transformation function, and $e \in \{0, 1\}$ is a parameter defining the encryption ($e = 0$) or decryption ($e = 1$) mode. The secret key represents the concatenation of four 32-bit subkeys K_i, $i = 1, 2, 3, 4$: $K = (K_1, K_2, K_3, K_4)$. The cipher uses simple key scheduling. The extended key $Q^{(e)} = (Q_1^{(e)}, Q_2^{(e)}, \ldots, Q_{10}^{(e)})$, where $Q_j^{(e)}$ represents the round keys ($j = 1, 2, \ldots, 10$), is formed as some sequence of subkeys, K_i. DDP-64's encryption round **Crypt**$^{(e)}$ is shown in Figure 2.11*a*. The data ciphering procedure is described as follows:

Table 2.8 Scheduling for the round subkeys and switching bit e′

$j =$	1	2	3	4	5	6	7	8	9	10
$A^{(1)} =$	Q_3	Q_2	Q_1	Q_4	Q_3	Q_3	Q_4	Q_1	Q_2	Q_3
$A^{(2)} =$	Q_4	Q_3	Q_2	Q_1	Q_2	Q_2	Q_1	Q_2	Q_3	Q_4
$A^{(3)} =$	Q_1	Q_4	Q_3	Q_2	Q_1	Q_1	Q_2	Q_3	Q_4	Q_1
$A^{(4)} =$	Q_2	Q_1	Q_4	Q_3	Q_4	Q_4	Q_3	Q_4	Q_1	Q_2
$e'^{(e=0)}$	1	0	1	1	0	1	1	1	0	1
$e'^{(e=1)}$	0	1	0	0	0	1	0	0	1	0

1. Perform initial transformation of the input data block $X = (L', R')$: $(L_0, R_0) = (L' \oplus Q_2, R' \oplus Q_1)$.
2. Perform ten data ciphering rounds using the procedure Crypt$^{(e)}$:
   2.1. For $j = 1$ to 9, execute transformation: $(L_j, R_j) = \mathbf{Crypt}^{(e)}(L_{j-1}, R_{j-1}, Q_j^{(e)})$. Swap the data subblocks: $T = R_j$, $R_j := L_j$, $L_j :=$ T.
   2.2. Execute transformation $(L_{10}, R_{10}) = \mathbf{Crypt}^{(e)}(L_9, R_9, Q_{10}^{(e)})$.
3. Perform final transformation: $Y = (L'', R'') = (L_{10} \oplus Q_4, R_{10} \oplus Q_3)$.

The key for each round, $Q_j^{(e)}$, contains four e-dependent subkeys $A^{(k)}$, $k = 1, 2, 3, 4$: $Q_j^{(e)} = (A^{(1)}, A^{(2)}, A^{(3)}, A^{(4)})^{(e)}{}_j$. The left data subblock L combined with subkey $A^{(1)}$ is used to form the controlling vector V, which specifies the current modification of the DDP operation performed on the right data subblock with the box $\mathbf{P}_{32/96}$. Then the left data subblock is transformed with $\Pi^{(e')}$ and rotation operations: $L \rightarrow L'' \rightarrow L'$. The L' subblock combined with subkey $A^{(3)}$ is used to form the controlling vector V', which specifies the current DDP modification performed on the right data subblock with the box $\mathbf{P}^{-1}_{32/96}$. The L and L' values combined with subkeys $A^{(2)}$ and $A^{(4)}$ are transformed with two **F** boxes implementing nonlinear DDP operations. Figure 2.11*b* and Table 2.8 specify the round subkeys and their correspondence to the secret key.

Changing the procedure from encryption to decryption is performed by simply swapping subkeys K_i with two single-layer boxes $\mathbf{P}^{(e)}_{64/1}$, as in the Cobra-H64 cipher, and switching the $\Pi^{(e')}$ operation in each transformation round, except the fifth and the sixth. The e'-dependent fixed permutation $\Pi^{(e')}$ in the left branch of the DDP-64 transformation round serves to prevent the homogeneity of the encryption procedure in case of the secret keys comprising four equal subkeys $K_1 = K_2 = K_3 = K_4$. This switchable operation allows one to avoid some hypothetical variants of the slide attacks [7], if some nonperiodic schedule of the switching bit e' is assigned, as specified for DDP-64 in Table 2.8. The design of switchable operations $\Pi^{(e')}$ is presented in Figure 2.11*c*, where $\Pi = \Pi^{(0)}$ and $\Pi^{-1} = \Pi^{(1)}$ are mutually inverse fixed permutations. The $\Pi^{(0)}$ bit permutation is described as follows:

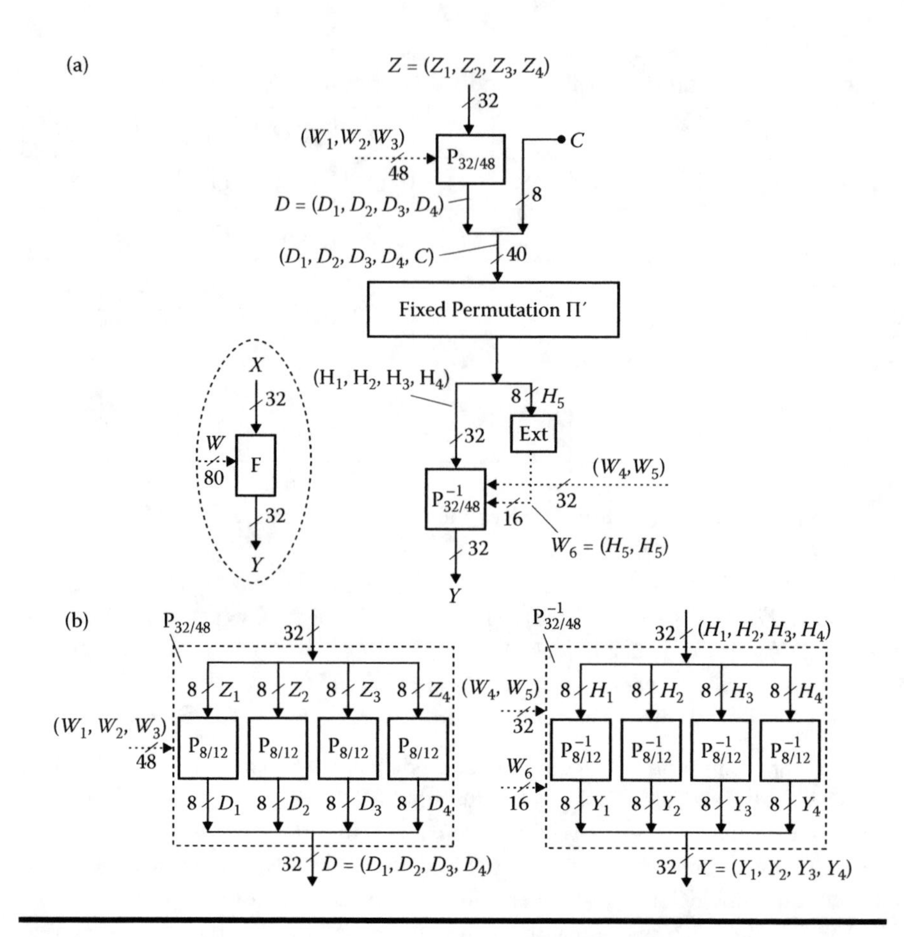

Figure 2.12 Design of the F box (*a*) and structure of the CP boxes $\mathbf{P}_{32/48}$ and $\mathbf{P}^{-1}_{32/48}$ (*b*).

(1, 4, 7, 2, 5, 8, 3, 6) (9, 12, 15, 10, 13, 16, 11, 14)
(17, 20, 23, 18, 21, 24, 19, 22)(25, 28, 31, 26, 29, 32, 27, 30)

The $\Pi^{(e')}$ operation is controlled with the switching bit e', depending on the switching bit e.

Variable permutations on the right data subblock are performed with the CP boxes $\mathbf{P}^{(V)}_{32/96}$ and $(\mathbf{P}^{-1}_{32/96})^{(V')}$ of the second order having the same topology as in the case of Cobra-H64. Controlling vectors V and V' are formed using the extension box **E** described as follows. Suppose the vector $L = (L_l, L_h)$ is fed to the **E**-box input. Then its output $V = (V_1, V_2, V_3, V_4, V_5, V_6)$ contains the following six 16-bit vectors:

$$V_1 = L_l,\ V_2 = L_l^{<<<6},\ V_3 = L_l^{<<<12},\ V_4 = L_h,\ V_5 = L_h^{<<<6},\ V_6 = L_h^{<<<12} \tag{2.28}$$

Nonlinear DDPs are performed with the **F** boxes that represent a compression function constructed using CP boxes and fixed permutations. The design of an **F** box is presented in Figure 2.12. The **F** box provides arbitrary change of the output vector weight. Indeed, depending on L, $A^{(2)}$, and $A^{(4)}$, eight of thirty-two input bits are replaced by the bits of the constant $C = (10101010)$ during the operation **F**. The **F** box comprises two three-layer CP boxes, $\mathbf{P}_{32/48}$ and $\mathbf{P}^{-1}_{32/48}$, separated with fixed permutation Π' described as follows:

(1, 33) (2, 9) (3, 17) (4, 25)(5)(6,13)(7,21)(8,34,29,40)(10,35)(11,18)
(12,26)(14)(15,36,22,38)(16,30)(19,37)(20,27)(23)(24,31)(28,39)(32)

The 80-bit controlling vector $W = (W_1, W_2, W_3, W_4, W_5)$, where $W_1, W_2, ..., W_5 \in \{0, 1\}^{16}$, of the **F** box is divided into two parts: a 48-bit controlling vector (W_1, W_2, W_3) of the $\mathbf{P}_{32/48}$ box and a 32-bit part (W_4, W_5) of the controlling vector of the $\mathbf{P}^{-1}_{32/48}$ box. The 16-bit vector W_6 is formed with the extension box "Ext" (Figure 2.12*a*) using the H_5 element of the output $H = (H_1, H_2, H_3, H_4, H_5)$, where $H_1, H_2, ..., H_5 \in \{0, 1\}^8$, of the Π' permutation. The output of the "Ext" box is the vector $W_6 = (H_5, H_5)$. The 80-bit controlling vector W is formed with the extension box $\mathbf{E}'$, the input of which is the vector $Z' = (Z'_l, Z'_h)$. The $\mathbf{E}'$ box is described by the following relation between Z' and W:

$$W_1 = Z'_l;\ W_2 = Z_l'^{<<<5};\ W_3 = Z_l'^{<<<10};\ W_4 = Z'_h;\ W_5 = Z_h'^{<<<5} \tag{2.29}$$

The vectors W_1, W_2, and W_3 control the first, second, and third active layers of the $\mathbf{P}_{32/48}$ box and vectors W_4, W_5, and W_6 control the corresponding active layers of the $\mathbf{P}^{-1}_{32/48}$ box. The output vector $D = (D_1, D_2, D_3, D_4) = (d_1, d_2, \ldots, d_{32})$ of the $\mathbf{P}_{32/48}$ box is concatenated with the constant C forming the vector (D_1, D_2, D_3, D_4, C), which is fed to the input of the fixed permutation Π'. The output vector of the Π' permutation is $(H_1, H_2, H_3, H_4, H_5)$, where $H_5 = (d_1, d_8, d_{10}, d_{15}, d_{19}, d_{22}, d_{28}, d_{29})$. Taking into account the structure of the $\mathbf{P}_{32/48}$ box, one can see that superposition $\mathbf{P}_{32/48}$ $^\circ\Pi'$ moves two arbitrary bits of each byte Z_i of the vector $Z = (Z_1, Z_2, Z_3, Z_4)$ to H_5 with the same probability. A single arbitrary bit of each byte Z_i moves to H_5 with probability 2^{-2}. Thus, the vector H_5 is composed of 8 bits of $Z = L \oplus A^{(2)}$, which, at the output of the Π' permutation, are replaced by 8 bits of C. Depending on W, the oddness of the output vector of the **F** box changes arbitrarily.

The "<<<16" rotation operation performed on the left data subblock is used to provide "symmetric" use of the most significant (L_h) and least significant (L_l) halves of L during the two F-box operations performed. The structure of the switchable permutation $\Pi^{(e')}$ provides the condition $(\Pi^{(e')}(L))^{<<<16} = \Pi^{(e')}(L^{<<<16})$ for $e' \in \{0, 1\}$, which is necessary for correct decryption.

The DDP-64 encryption algorithm presents an example of the pure DDP-based ciphers in which the CP boxes are extensively used in three different ways: (1) to implement bijective linear transformations, (2) to implement nonbijective nonlinear transformations, and (3) to implement switchable operation Π' and switchable key scheduling. Similar to the DDP-based cipher Cobra-H64, the cryptosystem DDP-64 is fast in case of frequent change of keys, because it is free of the key preprocessing. The avalanche effect spreads mostly when the changed bits are used as controlling ones, but not when they are transformed with the CP-box operations (some avalanche connected with the **F** boxes is defined by the use of eight input bits as an internal controlling vector, denoted earlier as W_6).

2.5.2 Security Estimation

The avalanche effect corresponding to the operations $\mathbf{P}_{32/96}$ or $\mathbf{P}^{-1}_{32/96}$ is caused by the use of the data subblock L to define the controlling vectors V and V', exactly as in the Cobra-H64 cipher. The avalanche effect corresponding to the **F** operations relates to the use of the left data subblock to specify the controlling vectors W and W'. Besides, the avalanche spreads due to the dependence of the output of the "Ext" box on L. Let us consider the vector $L = (L_l, L_h)$ before the operation "<<<16." Each bit l_i, where $1 \le i \le 16$, of L_l influences three elementary boxes $\mathbf{P}_{2/1}$ of the $\mathbf{P}_{32/48}$ box in the lower **F** box and two boxes $\mathbf{P}_{2/1}$ of the $\mathbf{P}^{-1}_{32/48}$ box in the upper **F** box. Besides, with probability 2^{-2} (this probability corresponds to the event that bit l_i is moved to H_5), the bit l_i influences two boxes $\mathbf{P}_{2/1}$ of the $\mathbf{P}^{-1}_{32/48}$ box in the upper **F** box and, with the same probability bit l_i, influences two boxes $\mathbf{P}_{2/1}$ of the $\mathbf{P}^{-1}_{32/48}$ box in the lower **F** box. All bits of L_h possess the same properties, because after the operation "<<<16" we have $(L_l, L_h)^{<<<16} = (L_h, L_l)$.

Differential analysis is the most powerful known attack on DDP-64. The best known DCs for DDP-64 correspond to the differences with a few active bits. The iterative two-round DCs with differences (Δ^L_1, Δ^R_0) and (Δ^L_0, Δ^R_1) have the highest probability: $P(2) \approx 1.37 \cdot 2^{-17}$. The difference (Δ^L_1, Δ^R_0) passes through eight and ten rounds of DDP-64 with probabilities $P(8) = P^4(2) \approx 1.79 \cdot 2^{-67}$ and $P(10) = P^5(2) \approx 1.23 \cdot 2^{-83}$, respectively. Thus, DDP-64 with eight and ten rounds is indistinguishable from a random cipher with differential attack using the most efficient two-round iterative characteristic.

Let us consider the formation mechanism of the iterative two-round DC with the difference (Δ^L_1, Δ^R_0). Passing through the left branch of the round transformation, the difference Δ^L_1 containing one active bit can produce zero difference at the output of the both **F** boxes. This takes place in the following two cases contributing mostly to the probability of the considered DC. Case 1 is connected with the following elementary events:

1. In both the **F** boxes, the active bit moves to one of the eight digits corresponding to the vector H_5 at the output of the Π' permutation (the probability of this event is $p_1 = 2^{-2} \cdot 2^{-2} = 2^{-4}$).
2. In both the **F** boxes, the active bit generates no pairs of active bits in the boxes $\mathbf{P}_{32/48}$ and $\mathbf{P}^{-1}_{32/48}$ (the probability of this event is $p_2 = 2^{-2} \cdot 2^{-3} \cdot (2^{-2})^2 = 2^{-9}$).

3. In box $\mathbf{P}_{32/96,}$ the active bit of the difference Δ_1^L generates no active bits of the right data subblock (the probability of this event is $p_3 = 2^{-3}$).
4. In box $\mathbf{P}^{-1}_{32/96,}$ the active bit of the difference Δ_1^L generates no active bits of the right data subblock (the probability of this event is $p_4 = 2^{-3}$).

In Case 2 we have the following elementary events:

1. In both the **F** boxes, the active bit moves to one of the thirty-two digits of the vector (H_1, H_2, H_3, H_4) at the output of the Π' permutation (the probability of this event is $p'_1 = (1 - 2^{-2})^2 \approx 1.12 \cdot 2^{-1}$).
2. In both the **F** boxes, the active bit generates no pairs of active bits in the boxes $\mathbf{P}_{32/48}$ and $\mathbf{P}^{-1}_{32/48}$ (the probability of this event is $p'_2 = 2^{-2} \cdot 2^{-3} = 2^{-5}$).
3. In box $\mathbf{P}_{32/96,}$ the active bit of the difference Δ_1^L generates no active bits of the right data subblock (the probability of this event is $p'_3 = 2^{-3}$).
4. In box $\mathbf{P}^{-1}_{32/96,}$ the active bit of the difference Δ_1^L generates no active bits of the right data subblock (the probability of this event is $p'_4 = 2^{-3}$).
5. At the output of the upper and lower **F** boxes are formed the differences $\Delta'_{1|i}$ and $\Delta'_{1|j}$, where $i = \mathbf{I}^*(j)$ (the probability of this event is $p'_5 = 2^{-5}$).

Denoting the probabilities of Cases 1 and 2 as P' and P'', respectively, we have

$$P' = p_1 p_2 p_3 p_4 = 2^{-4} \cdot 2^{-9} \cdot 2^{-3} \cdot 2^{-3} = 2^{-19} \tag{2.30}$$

and

$$P'' = p'_1 p'_2 p'_3 p'_4 p'_5 = 1.12 \cdot 2^{-1} \cdot 2^{-5} \cdot 2^{-3} \cdot 2^{-3} \cdot 2^{-5} = 1.12 \cdot 2^{-17} \tag{2.31}$$

There exist several other mechanisms for the formation of zero difference at the output of the right branch. However, their joint contribution is significantly smaller than the P' and P'' values. Thus, the difference (Δ_1^L, Δ_0^R) passes through one round with probability $P = P' + P'' = 1.37 \cdot 2^{-17}$. After swapping data subblocks the difference (Δ_0^L, Δ_1^R) is fed to input of the second round. The difference Δ_1^R passes through the right branch with probability 1 and, after swapping subblocks, we have the difference (Δ_1^L, Δ_0^R). We have the following probability of the iterative two-round DC $P(2) = P \approx 1.37 \cdot 2^{-17}$. Note that we also have the same probability for the two-round difference (Δ_0^L, Δ_1^R).

A linear attack can also be applied to distinguish the DDP-64 algorithm with reduced number of rounds from the random transformation. The best known LC comprises masks with two active bits. The formation mechanism of this LC takes into account that the bits of transformed data can be replaced by the bits of the constant $C = (10101010)$. Let $\mathbf{M} = (\mathbf{M}^L, \mathbf{M}^R)$ and $\mathbf{M}' = (\mathbf{M}'^L, \mathbf{M}'^R)$ are the input and output masks, respectively. Because of nonlinear operations performed with two **F** boxes, the masks $\mathbf{M} = \mathbf{M}' = (111\ldots1)$ have low bias and are not efficient. For each of the three individual boxes $\mathbf{P}_{32/96}$, $\mathbf{P}^{-1}_{32/96}$, and **F**, the LCs including masks with $t \leq 31$ active bits have a bias $b \leq 2^{-6}$. The maximum bias $b = 2^{-6}$ corresponds to the case $t = 1$. Calculation of the bias of CP boxes can be done, with the active bits' passing through the CP boxes being considered. Because the DDP-64 cipher represents a large permutation network, the mentioned approach can be applied to DDP-64.

Let us consider a one-round LC with the input $\mathbf{M} = \mathbf{M}^R_{1|j}$ and output $\mathbf{M}' = \mathbf{M}'^R_{1|g}$ masks. Thus, the bias of the sum $\sigma = r_j \oplus r_g$ is to be considered. The r_j active bit is moved to the arbitrarily given digit at the output of the $\mathbf{P}_{32/96}$ operation, with the probability $p_1 = 2^{-5}$. Therefore, it can be added with the ith bit of the left data subblock (the probability of this event is $0.75 \cdot 2^{-5}$ for all $i = 1, 2, \ldots, 32$) or with a bit of the constant C (the probability of this event is 0.25), while the output of the $\mathbf{P}_{32/96}$ operation is added to the output of the upper box $\mathbf{F}$. While the right data subblock is XORed with the output of the lower box $\mathbf{F}$, another given bit of L is added to the active bit of R with the probability $0.75 \cdot 2^{-5}$. As we do not take into account any bit of the left data subblock, the σ value is equal to 1 with probability 0.5, if two independent bits of L are added to the active bit of R or at least one bit of constant is added to the active bit. The nonzero bias of the considered LC is defined by the probability that two dependent bits of L are added to the active bit of the right data subblock (see Figure 2.13). For all values $i = 1, 2, \ldots, 32$, the bits l_i at the input of the $\Pi^{(e')}$ operation and $l_{i'}$ at the output of the $\Pi^{(e')}$ operation are dependent, if $i' = \Pi^{(e')}(i)$. Indeed, in the instance just mentioned, we have $l_{i'} = l_i$. Thus, at the input of the $\mathbf{P}^{-1}_{32/96}$ box, the value of the active bit is equal to $r_{z'} = r_j \oplus l_i \oplus a'_i \oplus l_{i'} \oplus a''_{i'} = r_j \oplus a'_i \oplus a''_{i'}$, with the probability $(0.75 \cdot 2^{-5})^2$. The r_x active bit is moved to the bit r_g at the output of the $\mathbf{P}^{-1}_{32/96}$ box with the probability 2^{-5}; i.e., with the probability $p'_i = (0.75 \cdot 2^{-5})^2 \cdot 2^{-5} = 0.56 \cdot 2^{-15}$, we have $\sigma = \sigma_i = r_j \oplus r_g = a'_i \oplus a''_{i'}$. If the active bit is moved to bit $r_{g'}$, where $g' \neq g$, then $\Pr(\sigma = 1) = 0.5$, as in this case the r_g value is unbiased. We have 2^5 independent events that contribute to the bias of the considered LC.

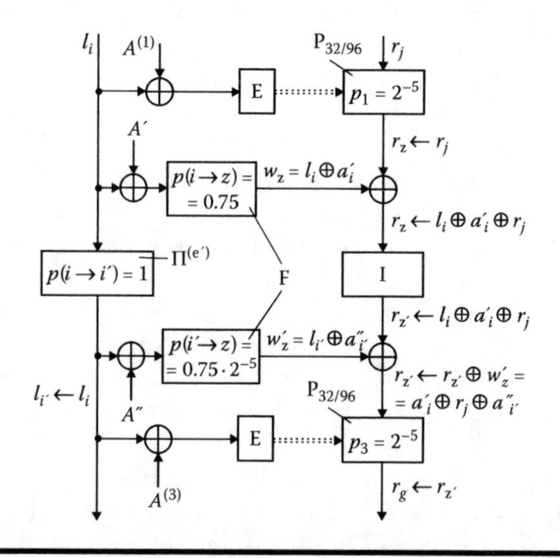

Figure 2.13 Formation of the one-round linear characteristic $[(\mathrm{M}^L_0, \mathrm{M}^R_{1|j}); (\mathrm{M}'^L_0, \mathrm{M}'^R_{1|g}); b(1)]$ in DDP-64.

For all $i \in \{1, 2, ..., 32\}$ the probability p_i' is equal to $p_1' = 0.56 \cdot 2^{-15}$. The bias value of the considered LC can be calculated using the following formula:

$$b(1) = \left| \Pr(\sigma = 1) - \frac{1}{2} \right| = \left| \frac{1}{2}\left(1 - 2^5 p_1'\right) + p_1' N_1 - \frac{1}{2} \right| \tag{2.32}$$

where $N_1 = \#\{i\text{: } a_i' \oplus a_{i'}'' = 1\}$. Assuming that the bits of the subkeys A' and A'' are uniform random variables, we get $\langle N_1 \rangle = 2^4$ and $N_1 = 2^4 + d$, where d is the deviation of the random value N_1 from its average ($-2^4 \le d \le 2^4$). Using this representation of the N_1 value, we get:

$$b(1) = |p'd| \le 0.56 \cdot 2^{-11} \tag{2.33}$$

As in the second round the active bit goes through the left branch of the round transformation, its value does not change and its position at the output of the $\Pi^{(e')}$ operation is predetermined. Therefore, the bias of the LC ($\mathbf{M}^R_{1|j}$, $\mathbf{M'}^R_{1|g}$, $b(2)$) corresponding to the transformation **Crypt°T°Crypt°T** representing the superposition of two round transformations and two transposition operations is equal to the value $b(1)$; i.e., we have $b(2) = b(1)$. The bias of the LC ($\mathbf{M}^R_{1|j}$, $\mathbf{M'}^R_{1|g}$, $b(2k)$) corresponding to the k pairs of the round transformations, followed by the **T** operation, is defined by the events including the cases of adding dependent bits to the active bit in each odd round and moving the active bit to the bit $r_{g'}$, where $g' = \Pi^{(e'\oplus 1)}(g)$, at the output of the $\mathbf{P}^{-1}_{32/96}$ operation in the $(2k - 1)$th round. In this case we have

$$\sigma = r_j \oplus r_g = \sum_z a'_{i_z} \oplus a''_{i'_z}, \qquad z \in \{1, 3, ..., 2k-1\} \tag{2.34}$$

where $i_z' = \Pi^{(e')}(i_z)$ for all values z. As in the case of the one-round LC, we can write

$$b(2k) = \left| \Pr(\sigma = 1) - \frac{1}{2} \right| = \left| \frac{1}{2}\left(1 - 2^{5k} p_k'\right) + p_k' N_k - \frac{1}{2} \right|, \tag{2.35}$$

where

$$p_k' = (3/4)^{-2k} \cdot \left(2^{-10}\right)^k \cdot 2^{-5} \approx 2^{-11k-4} \text{ and } N_k = \#\left\{ (i_1, i_3, ..., i_z) : \sum_z a'_{i_z} \oplus a''_{i'_z} = 1 \right\}.$$

Using the representation $N_k = \langle N_k \rangle + d_k$, where d_k ($-2^{5k-1} \le d_k \le 2^{5k-1}$) is the deviation of the random value N_k from its average $\langle N_k \rangle = 2^{5k-1}$, we get

$$b(2k) = |p_k' d_k| \le 2^{-6k-5} \tag{2.36}$$

The average value of d_k can be estimated as $\langle d_k \rangle \approx \sqrt{2^{5k}} = 2^{2.5k}$; therefore, for some "average" key, we have the bias value $b(2k) \approx 2^{-8.5k-4}$. For the random cipher LCs have the bias $b = 2^{-32} >> b(2k)$, if $k = 4, 5$. This comparison shows that eight rounds of DDP-64 are sufficient to thwart linear attacks. Equation (2.36) shows that ten rounds are sufficient to thwart LCA in all cases, including improbable cases that correspond to the maximum value $|d_k| = 2^{5k-1}$.

The constant $C = (10101010)$ has been selected to prevent the mechanism of formation of the LC $[(\mathbf{M}_0^L, \mathbf{M}_{1|j}^R); (\mathbf{M}'^L_0, \mathbf{M}'^R_{1|g}); b'(2)]$, which is connected with the events of XORing the active bits passing through the right branch with randomly selected bits of the constant C. This mechanism defines the following value of the $b'(2)$ bias:

$$b'(2) = 2^{-5}\left[-\frac{1}{32} + \frac{1}{16}\cdot\frac{\varphi^2(C) + (8 - \varphi(C))^2}{64}\right] \tag{2.37}$$

For the used value C, we have $\varphi(C) = 4$; therefore $b'(2) = 0$. The worst case of selection of the C value corresponds to $\varphi(C) = 8$ or $\varphi(C) = 0$. In the worst case of the C value, we have the bias value $b'(2k) = 2^{-6-4k}$ and 14 rounds of DDP-64 are required to thwart LCA.

In spite of the simplicity of the key schedule, the "symmetric" keys such as $K' = (X, Y, X, Y)$ and $K'' = (X, X, X, X)$ are not weak or semiweak, because decryption requires switching the fixed permutation in the left branch of the DDP-64 encryption round. Slide attacks in the case of "symmetric" keys are inefficient, because encryption with DDP-64 is free of homogeneity (in the sense used by Biryukov et al. [7]) due to the nonperiodic schedule of the switching bit e' specifying the fixed permutation $\Pi^{(e')}$ performed on the left data subblock. This shows that the switchable operations can play a sufficiently important role in block ciphers that are free of key preprocessing. For comparison, one can remark that SPECTR-H64, which uses no switchable operations, has weak keys (for all X its 256-bit key $K'' = (X, X, \ldots, X)$ is a weak one) and, in their case, seems to be vulnerable to slide attack.

The use of some secret key preprocessing is a standard way to prevent weak keys and homogeneity in DDP-64. However, this significantly increases the hardware implementation cost. The use of switchable operations represents an interesting alternative affording reduction in implementation cost and providing faster encryption in case of frequent change of keys.

2.6 Conclusions

- Theoretical analysis of the LCs of DDP operations conserving the weight of the transformed vector has shown that the principal problem in the design of DDP-based ciphers is the need to prevent LCAs from using the masks of maximum weight.

- There are two main ways of using DDPs as cryptographic primitives: (1) implementing bijective linear transformations and (2) implementing nonlinear compression transformations.
- Switchable (e-dependent) operations are a new interesting primitive in the design of block ciphers using simple key scheduling.
- Differential analysis seems to be more efficient against the DDP-based ciphers than linear analysis.
- Differential and linear characteristics with fewer active bits seem to be, respectively, more efficient than DCs and LCs with larger numbers of active bits.

Chapter 3

Data-Driven Primitives and Ciphers Based on Controlled Substitution–Permutation Networks

This chapter introduces some generalizations of the DDP-based approach to the synthesis of the block ciphers based on data-dependent operations (DDOs). Permutation networks are presented as a particular case of the controlled substitution–permutation networks (CSPNs) with minimum size controlled elements (CEs). It is shown that CSPNs allow one to design data-driven primitives (DDPs) that have more advanced properties than the variable permutations. A classification of the minimum size CEs is proposed. The properties of different DDOs are described. Symmetric topologies for designing the DDOs are described. The next step of advancing the design of the DDOs suitable to cheap hardware implementation is the use of the CEs with two-bit controlled input. Selection criteria and classification of such CEs are presented. Several ciphers based on CSPNs are described and estimated.

3.1 Advanced DDP-Like Primitives and Their Classification

3.1.1 Elementary Controlled Substitutions

The main building block in the layered controlled permutation (CP) boxes is the elementary switching element $\mathbf{P}_{2/1}$ performing controlled transposition of two input bits. The elementary controlled transformation performed with $\mathbf{P}_{2/1}$ is described by two specific nonlinear Boolean functions (BFs) in three variables:

$$y_1 = f_1(x_1, x_2) = x_1 v \oplus x_2 v \oplus x_1 \quad (3.1)$$

$$y_2 = f_2(x_1, x_2) = x_1 v \oplus x_2 v \oplus x_2 \quad (3.2)$$

where v is the controlling bit. Selecting two balanced BFs in three variables f_1 and f_2 of different types, one can get different variants of the elementary controlled boxes $\mathbf{F}_{2/1}$ (see Figure 3.1). The f_1 and f_2 functions can be used when implementing the $\mathbf{F}_{2/1}$ boxes in hardware. Controlled elements $\mathbf{F}_{2/1}$ can be alternatively described as two different linear substitutions S_1 (if $v = 0$) and S_2 (if $v = 1$) of the 2×2 size (Figure 3.1*c*). The relation between the mentioned representations of the CEs

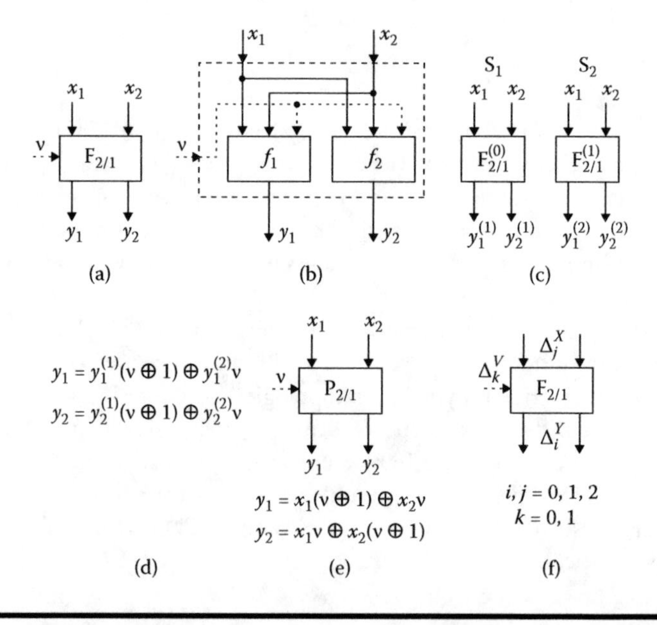

Figure 3.1 Elementary box $F_{2/1}$: (*a*) general case; (*b*) implementation; (*c*) representation as a pair of two 2 × 2 substitutions; (*d*) formulas describing relation between representations *b* and *c*; (e) Boolean functions describing the box $P_{2/1}$; (*f*) differences corresponding to the differential characteristics of the $F_{2/1}$ elements.

is given by the formulas presented in Figure 3.1*d*. Outputs of the S_1 box can be described by a pair of BFs in two variables:

$$y_1^{(1)} = f_1^{(1)}(x_1, x_2) \text{ and } y_2^{(1)} = f_2^{(1)}(x_1, x_2) \tag{3.3}$$

The outputs of the S_2 box are described respectively, as follows:

$$y_1^{(2)} = f_1^{(2)}(x_1, x_2) \text{ and } y_2^{(2)} = f_2^{(2)}(x_1, x_2) \tag{3.4}$$

The functions $f_1^{(1)}, f_2^{(1)}, f_1^{(2)}$, and $f_2^{(2)}$ define a pair of BFs in three variables, which describe some CEs as a whole. If $v = 0$, then CE implements the S_1 box substitution. If $v = 1$, then CE implements the S_2 box substitution. In other words the controlling bit defines selection of the current elementary substitution operation ($\mathbf{F}_{2/1}^{(0)}$ or $\mathbf{F}_{2/1}^{(1)}$ modification of the $\mathbf{F}_{2/1}$ operation). The formulas shown in Figure 3.1*d* describe the selection of the current modification in some evident form. The formulas can be rewritten as follows:

$$y_1 = (v \oplus 1) f_1^{(1)} \oplus v f_1^{(2)} = v(y_1^{(1)} \oplus y_1^{(2)}) \oplus y_1^{(1)} \tag{3.5}$$

$$y_2 = (v \oplus 1) f_2^{(1)} \oplus v f_2^{(2)} = v(y_2^{(1)} \oplus y_2^{(2)}) \oplus y_2^{(1)} \tag{3.6}$$

Using some given topology of the $\mathbf{P}_{n/m}$ box and replacing the $\mathbf{P}_{2/1}$ units by $\mathbf{F}_{2/1}$ elements of different types, we can get different variants of the controlled operational boxes $\mathbf{F}_{n/m}$ performing transformations of different types, i.e., those that in a general case do not conserve the weight of the transformed binary vectors. A heterogeneous box $\mathbf{F}_{n/m}$ can be composed using elementary boxes $\mathbf{F}_{2/1}$ of several different types, for example, each active layer can be unique. Usually we consider the $\mathbf{F}_{n/m}$ boxes with uniform structure, which are built up using elementary boxes $\mathbf{F}_{2/1}$ of the single type, and nonuniform boxes constructed using two types of the CEs that represent mutual inverses $\mathbf{F}_{2/1}$ and $\mathbf{F}_{2/1}^{-1}$. The $\mathbf{F}_{n/m}$ boxes represent different types of the CSPNs built up using minimum size CEs.

In many cases, the use of the operation $\mathbf{F}_{n/m}$ while encrypting implies the use of its inverse $\mathbf{F}_{n/m}^{-1}$. It is evident that arbitrary $\mathbf{F}_{n/m}$ box operation is invertible if the $\mathbf{F}_{2/1}$ element is invertible. Inverse transformation can be constructed by swapping the input and output of the given $\mathbf{F}_{n/m}$ box and replacing each of the $\mathbf{F}_{2/1}$ elements by its inverse $\mathbf{F}_{2/1}^{-1}$. (Alternatively, by replacing the $\mathbf{P}_{2/1}$ elements with $\mathbf{F}_{2/1}^{-1}$ ones in the $\mathbf{P}_{n/m}^{-1}$ box topology, we get the $\mathbf{F}_{n/m}^{-1}$ box). To define the easy construction of the mutual inverse operations, one can use the $\mathbf{F}_{2/1}$ building units that are elementary controlled involutions (both modifications $\mathbf{F}_{2/1}^{(0)}$ and $\mathbf{F}_{2/1}^{(1)}$ of such elements are involutions). The box $\mathbf{P}_{2/1}$ is a particular type of the elementary controlled involution. In the following text, we show that there are 40 elementary controlled involutions and 24 of them are more interesting elementary cryptographic primitives than the $\mathbf{P}_{2/1}$ box.

3.1.2 Classification of the $F_{2/1}$ Boxes

To perform a classification of the $\mathbf{F}_{2/1}$ elements and select those that are more suitable for cryptographic applications, we need to choose some selection criteria. Using the notion of nonlinearity (NL) in the sense of distance from the set of affine BFs in the same number of variables, the following criteria have been chosen to select nonlinear elements $\mathbf{F}_{2/1}$ suitable for designing the $\mathbf{F}_{n/m}$ boxes for cryptographic applications [40,43,45]:

1. Each of two outputs of the CEs should be a nonlinear BF having maximum possible NL for balanced BFs in three variables.
2. Each modification of the CEs should be a bijective transformation $(x_1, x_2) \rightarrow (y_1, y_2)$.
3. The linear combination of two outputs of the CEs, i.e., $f = y_1 \oplus y_2$, should have maximum possible NL for balanced BFs in three variables.
4. Each modification of the CEs should be an involution.

Selecting two balanced BFs in three variables f_1 and f_2 of different types, one can get different variants of the elementary controlled boxes $\mathbf{F}_{2/1}$ (see Figure 3.1*b*). There are only 70 balanced BFs in three variables. (A BF containing in its truth table equal number of zero and nonzero bits is a balanced one). Our interest in the balanced BF is connected with simplifying the exhaustive search of all possible invertible CEs $\mathbf{F}_{2/1}$ that can be used in the design of invertible DDP-like operations. It is known that a transformation operation is bijective (i.e., invertible) if each of its output bits is described by a balanced BF and all linear combinations of output bits are also balanced BFs. Therefore, we need to consider only the pairs of different balanced BF number. The number of such pairs is $70 \cdot 69 = 4830$.

There are only 24 different S boxes of the 2×2 size. Therefore, to perform an exhaustive search, we need to consider only $24 \cdot 23 = 552$ variants. Thus, the last approach (which can be called visual design of the $\mathbf{F}_{2/1}$ elements) is more suitable than consideration of all possible pairs of the balanced BFs. To represent visually some CEs, one can describe them as pairs (S_1, S_2) of simplest transformations performing the 2×2 substitutions S_1 ($v = 0$) and S_2 ($v = 1$). The existing 24 different types of such S boxes can be represented as transformations shown in Figure 3.2. For example, the switching element $\mathbf{P}_{2/1}$ can be represented as the (a,e) element.

There are 288 CEs satisfying criteria 1 and 2. Criteria 1 to 3 define 192 different variants of the nonlinear CEs $\mathbf{F}_{2/1}$. Other 96 elements including the $\mathbf{P}_{2/1}$ element are linear primitives as the sum of their outputs $f = y_1 \oplus y_2$ is a linear BF. Criteria 1, 2, and 4 define 40 different variants of the CEs $\mathbf{F}_{2/1}$, 24 of them satisfying also criterion 3. These 40 types of CEs represent elementary controlled involutions.

Differential characteristics (DCs) of the $\mathbf{F}_{n/m}$ boxes are defined by their topology and DCs of the elementary controlled boxes used as main building blocks while constructing the $\mathbf{F}_{n/m}$ boxes. As a general case, the differences passing through the $\mathbf{F}_{2/1}$ element are shown in Figure 3.1*f*. Differential characteristics of the CEs can be

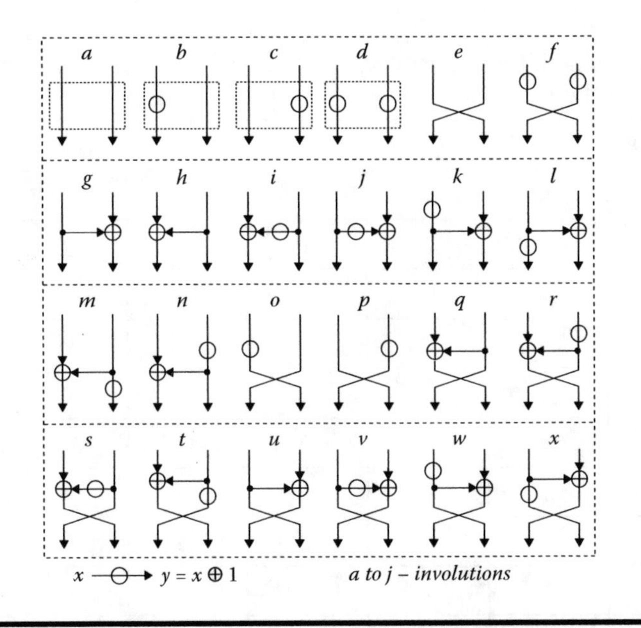

Figure 3.2 Visual representation of all possible types of the 2 × 2 S boxes.

divided into six groups (see Table 3.1). It is remarkable that we have only two types of nonlinear CEs, denoted as $\mathbf{Q}_{2/1}$ and $\mathbf{R}_{2/1}$ elements. The linear CEs ($\mathbf{Z}_{2/1}$) are divided into four groups: $\mathbf{Z}'_{2/1}$, $\mathbf{Z}''_{2/1}$, $\mathbf{Z}^{*}_{2/1}$, and $\mathbf{Z}^{\circ}_{2/1}$. One can see that the nonlinear CEs are less "predictable" than linear ones. If a CE belongs to the set of elements $\mathbf{Q}_{2/1}$ ($\mathbf{R}_{2/1}$), then its inverse also belongs to the same set. This shows that mutually inverse CEs of the $\mathbf{Q}_{2/1}$ and $\mathbf{R}_{2/1}$ types and the mutually inverse $\mathbf{F}_{n/m}$ and $\mathbf{F}^{-1}_{n/m}$ boxes based on them have analogous properties as cryptographic primitives. The elementary switching element $\mathbf{P}_{2/1}$ is an element of the subset $\mathbf{Z}^{\circ}_{2/1}$, i.e., $\mathbf{P}_{2/1} \in \{\mathbf{Z}^{\circ}_{2/1}\}$. Thus, the proposed classification defines the following properties:

- If $\mathbf{F}_{2/1} \in \{\mathbf{Q}_{2/1}\}$, then $\mathbf{F}^{-1}_{2/1} \in \{\mathbf{Q}_{2/1}\}$.
- If $\mathbf{F}_{2/1} \in \{\mathbf{R}_{2/1}\}$, then $\mathbf{F}^{-1}_{2/1} \in \{\mathbf{R}_{2/1}\}$.
- If $\mathbf{F}_{2/1} \in \{\mathbf{Z}^{*}_{2/1}\} \cup \{\mathbf{Z}^{\circ}_{2/1}\}$, then $\mathbf{F}^{-1}_{2/1} \in \{\mathbf{Z}^{*}_{2/1}\} \cup \{\mathbf{Z}^{\circ}_{2/1}\}$.
- If $\mathbf{F}_{2/1} \in \{\mathbf{Z}'_{2/1}\} \cup \{\mathbf{Z}''_{2/1}\}$, then $\mathbf{F}^{-1}_{2/1} \notin \{\mathbf{Z}_{2/1}\} \cup \{\mathbf{Q}_{2/1}\} \cup \{\mathbf{R}_{2/1}\}$.

All elementary controlled involutions $\mathbf{F}_{2/1}$ are presented in the upper left part of Table 3.2 that illustrates the classification of the CEs corresponding to the selection criteria used. Among 40 involutions $\mathbf{F}_{2/1}$, we have 8 involutions $\mathbf{Q}_{2/1}$, 16 involutions $\mathbf{R}_{2/1}$, and 16 involutions $\mathbf{Z}_{2/1}$. The last subset represents linear elementary controlled involutions among which we have two variants implementing the switching elements $\mathbf{P}_{2/1}$ and $\mathbf{P}'_{2/1}$, denoted as (a,e) and (e,a), respectively. Nonlinear controlled involutions $\mathbf{F}_{2/1}$ are described in Table 3.3.

Table 3.1 Probabilities $p_{ijk} = \Pr(\Delta_j^X \rightarrow \Delta_i^Y/\Delta_k^V)$ of differential characteristics of the $F_{2/1}$ elements

i	j	k	$\mathbf{R}_{2/1}$	$\mathbf{Q}_{2/1}$	$\mathbf{Z}'_{2/1}$	$\mathbf{Z}''_{2/1}$	$\mathbf{Z}^*_{2/1}$	$\mathbf{Z}^\circ_{2/1}$
0	0	1	¼	¼	0	½	0	½
1	0	1	½	½	1	0	1	0
2	0	1	¼	¼	0	½	0	½
0	1	1	¼	¼	¼	¼	½	0
1	1	1	½	½	½	½	0	1
2	1	1	¼	¼	¼	¼	½	0
1	1	0	¾	½	½	½	1	1
2	1	0	¼	½	½	½	0	0
1	2	0	½	1	1	1	0	0
2	2	0	½	0	0	0	1	1
0	2	1	¼	¼	½	0	0	½
1	2	1	½	½	0	1	1	0
2	2	1	¼	¼	½	0	0	½
Examples:			(e,g), (f,i), (p,h), (x,d)	(h,g), (j,i), (u,t), (x,r)	(l,u), (k,x), (n,s), (j,u)	(l,v), (l,s), (n,r), (j,x)	(b,e), (c,f), (a,o), (d,p)	(a,f), (e,a), (b,o), (c,p)

3.1.3 Subset of the $U_{2/1}$ Boxes with One Linear Output

Considering the $\mathbf{F}^{-1}_{2/1}$ boxes that are inverses of the nonlinear CEs $\mathbf{R}_{2/1}$, $\mathbf{Q}_{2/1}$, and $\mathbf{Z}_{2/1} \in \{\mathbf{Z}^*_{2/1}\} \cup \{\mathbf{Z}^\circ_{2/1}\}$, we have established that $\mathbf{F}^{-1}_{2/1}$ elements relate to the same type as $\mathbf{F}_{2/1}$. An interesting exception corresponds to the subset $\{\mathbf{Z}'_{2/1}\} \cup \{\mathbf{Z}''_{2/1}\}$. The inverses $\mathbf{F}'^{-1}_{2/1}$ of the elements $\mathbf{F}'_{2/1} \in \{\mathbf{Z}'_{2/1}\} \cup \{\mathbf{Z}''_{2/1}\}$ do not satisfy the selection criterion 1 that requires both outputs of the CE to have the maximum possible NL. The CEs are described by the balanced BFs f in three variables; therefore, only one NL value NL(f) is possible, i.e., NL(f) = 2. The last means that one output of the $\mathbf{F}'^{-1}_{2/1}$ elements is a linear or affine BF. (At least one output of $\mathbf{F}'^{-1}_{2/1}$ is a nonlinear BF. In the other case, the CEs $\mathbf{F}'_{2/1} = (\mathbf{F}'^{-1}_{2/1})^{-1}$ should define two linear outputs; however, we have selected the $\mathbf{F}'_{2/1}$ element from the subset satisfying criterion 1.) More detailed investigation of the elements $\mathbf{U}_{2/1} \in \{\mathbf{F}'^{-1}_{2/1} : \mathbf{F}'_{2/1} \in \{\mathbf{Z}'_{2/1}\}\}$ has shown that they represent specific interest for application in the synthesis of the controlled operations.

The peculiarity of the $\mathbf{Z}'_{2/1}$ elements is that the active bits of the differences at the controlling input influence deterministically the output difference. None of

Table 3.2 All existing controlled elements satisfying the selection criteria

	a	b	c	d	e	f	g	h	i	j	k	l	m	n	o	p	q	r	s	t	u	v	w	x
a	=				**P**	**Z**									**Z**	**Z**	**R**	**R**	**R**	**R**	**R**	**R**	**R**	**R**
b		=			**Z**	**Z**									**Z**	**Z**	**R**	**R**	**R**	**R**	**R**	**R**	**R**	**R**
c			=		**Z**	**Z**									**Z**	**Z**	**R**	**R**	**R**	**R**	**R**	**R**	**R**	**R**
d				=	**Z**	**Z**									**Z**	**Z**	**R**	**R**	**R**	**R**	**R**	**R**	**R**	**R**
e	**P**	**Z**	**Z**	**Z**	=		**R**	**R**	**R**	**R**	**R**	**R**	**R**	**R**										
f	**Z**	**Z**	**Z**	**Z**		=	**R**	**R**	**R**	**R**	**R**	**R**	**R**	**R**										
g					**R**	**R**	=	**Q**	**Q**				**Q**	**Q**	**R**	**R**					**Z**	**Z**	**Z**	**Z**
h					**R**	**R**	**Q**	=		**Q**	**Q**	**Q**			**R**	**R**	**Z**	**Z**	**Z**	**Z**				
i					**R**	**R**	**Q**		=	**Q**	**Q**	**Q**			**R**	**R**	**Z**	**Z**	**Z**	**Z**				
j					**R**	**R**		**Q**	**Q**	=			**Q**	**Q**	**R**	**R**					**Z**	**Z**	**Z**	**Z**
k					**R**	**R**		**Q**	**Q**		—		**Q**	**Q**	**R**	**R**					**Z**	**Z**	**Z**	**Z**
l					**R**	**R**		**Q**	**Q**			—	**Q**	**Q**	**R**	**R**					**Z**	**Z**	**Z**	**Z**
m					**R**	**R**	**Q**			**Q**	**Q**	**Q**	—		**R**	**R**	**Z**	**Z**	**Z**	**Z**				
n					**R**	**R**				**Q**	**Q**	**Q**		—	**R**	**R**	**Z**	**Z**	**Z**	**Z**				
o	**Z**	**Z**	**Z**	**Z**			**R**	**R**	**R**	**R**	**R**	**R**	**R**	**R**	—									
p	**Z**	**Z**	**Z**	**Z**			**R**	**R**	**R**	**R**	**R**	**R**	**R**	**R**		—								
q	**R**	**R**	**R**	**R**				**Z**	**Z**				**Z**	**Z**			—				**Q**	**Q**	**Q**	**Q**
r	**R**	**R**	**R**	**R**				**Z**	**Z**				**Z**	**Z**				—			**Q**	**Q**	**Q**	**Q**
s	**R**	**R**	**R**	**R**				**Z**	**Z**				**Z**	**Z**					—		**Q**	**Q**	**Q**	**Q**
t	**R**	**R**	**R**	**R**				**Z**	**Z**				**Z**	**Z**						—	**Q**	**Q**	**Q**	**Q**
u	**R**	**R**	**R**	**R**			**Z**			**Z**	**Z**	**Z**					**Q**	**Q**	**Q**	**Q**	—			
v	**R**	**R**	**R**	**R**			**Z**			**Z**	**Z**	**Z**					**Q**	**Q**	**Q**	**Q**		—		
w	**R**	**R**	**R**	**R**			**Z**			**Z**	**Z**	**Z**					**Q**	**Q**	**Q**	**Q**			—	
x	**R**	**R**	**R**	**R**			**Z**			**Z**	**Z**	**Z**					**Q**	**Q**	**Q**	**Q**				—

Table 3.3 The full set of nonlinear elementary controlled involutions

CE	Type	$f_1(x_1,x_2,v)$		$f_2(x_1,x_2,v)$	
		Algebraic normal form	Truth table	Algebraic normal form	Truth table
(g,e)	R	$x_1v\oplus x_2v\oplus x_1$	00011011	$x_2v\oplus x_1\oplus x_2$	00101101
(g,h)	Q	$x_2v\oplus x_1$	00011110	$x_1v\oplus x_1\oplus x_2$	00111001
(e,g)	R	$x_1v\oplus x_2v\oplus x_2$	00100111	$x_2v\oplus x_1$	00011110
(e,h)	R	$x_1v\oplus x_2$	00110110	$x_1v\oplus x_2v\oplus x_1$	00011011
(h,g)	Q	$x_2v\oplus x_1\oplus x_2$	00101101	$x_1v\oplus x_2$	00110110
(h,e)	R	$x_1v\oplus x_1\oplus x_2$	00111001	$x_1v\oplus x_2v\oplus x_2$	00100111
(g,i)	Q	$x_2v\oplus x_1\oplus v$	01001011	$x_1v\oplus x_1\oplus x_2$	00111001
(g,f)	R	$x_1v\oplus x_2v\oplus x_1\oplus v$	01001110	$x_2v\oplus x_2\oplus x_1\oplus v$	01111000
(i,g)	Q	$x_2v\oplus x_2\oplus x_1\oplus v\oplus 1$	10000111	$x_1v\oplus x_2$	00110110
(f,g)	R	$x_1v\oplus x_2v\oplus x_2\oplus v\oplus 1$	10001101	$x_2v\oplus x_1\oplus v\oplus 1$	10110100
(i,f)	R	$x_1v\oplus x_2\oplus x_1\oplus 1$	11000110	$x_1v\oplus x_2v\oplus x_2\oplus v$	01110010
(f,i)	R	$x_1v\oplus x_2\oplus 1$	11001001	$x_1v\oplus x_2v\oplus x_1\oplus v\oplus 1$	10110001
(h,j)	Q	$x_2v\oplus x_1\oplus x_2$	00101101	$x_1v\oplus x_2\oplus v$	01100011
(j,h)	Q	$x_2v\oplus x_1$	00011110	$x_1v\oplus x_2\oplus x_1\oplus v\oplus 1$	10010011
(j,f)	R	$x_1v\oplus x_2v\oplus x_1\oplus v$	01001110	$x_2v\oplus x_2\oplus x_1\oplus 1$	11010010
(f,h)	R	$x_1v\oplus x_2\oplus v\oplus 1$	10011100	$x_1v\oplus x_2v\oplus x_1\oplus v\oplus 1$	10110001
(f,j)	R	$x_1v\oplus x_2v\oplus x_2\oplus v\oplus 1$	10001101	$x_2v\oplus x_1\oplus 1$	11100001
(e,j)	R	$x_1v\oplus x_2v\oplus x_2$	00100111	$x_2v\oplus x_1\oplus v$	01001011
(j,e)	R	$x_1v\oplus x_2v\oplus x_1$	00011011	$x_2v\oplus x_2\oplus x_1\oplus v\oplus 1$	10000111
(j,i)	Q	$x_2v\oplus x_1\oplus v$	01001011	$x_1v\oplus x_2\oplus x_1\oplus v\oplus 1$	10010011
(i,e)	R	$x_1v\oplus x_2\oplus x_1\oplus v\oplus 1$	10010011	$x_1v\oplus x_2v\oplus x_2$	00100111
(i,j)	Q	$x_2v\oplus x_2\oplus x_1\oplus v\oplus 1$	10000111	$x_1v\oplus x_2\oplus v$	01100011
(h,f)	R	$x_1v\oplus x_1\oplus x_2\oplus v$	01101100	$x_1v\oplus x_2v\oplus x_2\oplus u$	01110010
(e,i)	R	$x_1v\oplus x_2\oplus v$	01100011	$x_1v\oplus x_2v\oplus x_1$	00011011

Table 3.4 Probabilities $p_{ijk} = \Pr(\Delta_j^X \to \Delta_i^Y/\Delta_k^V)$ of differential characteristics of the $U_{2/1}$ elements and the set $\{U_{2/1}\}$

$i =$	0	1	2	0	1	2	1	2	1	2	0	1	2
$j =$	0	0	0	1	1	1	1	1	2	2	2	2	2
$k =$	1	1	1	1	1	1	0	0	0	0	1	1	1
$p_{ijk} =$	**0**	½	½	½	½	**0**	½	½	**1**	**0**	**0**	½	½
$\{U_{2/1}\}$:	(g,r), (g,s), (h,v), (h,w), (i,u), (i,x), (j,q), (j,t), (k,q), (k,t), (l,r), (l,s),(m,v), (m,w), (n,u), (n,x), (r,g), (s,g), (v,h), (w,h), (u,i), (x,i), (q,j),(t,j), (q,k), (t,k), (r,l), (s,l), (v,m), (w,m), (u,n), (x,n)												

the elements $\mathbf{F}''_{2/1} \in \{\mathbf{Q}_{2/1}\} \cup \{\mathbf{R}_{2/1}\} \cup \{\mathbf{Z}''_{2/1}\} \cup \{\mathbf{Z}^{\circ}_{2/1}\}$ possesses such property (the $\mathbf{Z}^{*}_{2/1}$ elements possess such property, but they are less interesting because of their pure DCs). Therefore, with some probability, the CSPNs based on the elements $\mathbf{F}''_{2/1}$ conserve their output difference while a nonzero difference is fed to the controlling input. Considering the differential analysis of the algorithms Cobra-H64 and DDP-64, we have seen how this fact can be used to find DCs with high probabilities. Thus, some designs of block ciphers can be more efficient when using the CSPNs based on the $\mathbf{Z}'_{2/1}$ elements, imparting to the CSPN-based operations the property of deterministic change of the output difference when nonzero difference is fed to the controlling input. In different constructions of the iterated block ciphers, mutually inverse boxes $\mathbf{F}_{n/m}$ and $\mathbf{F}^{-1}_{n/m}$ are used; therefore, in such designs using the CSPNs based on the $\mathbf{Z}'_{2/1}$ elements, we should also use the CSPNs based on the $\mathbf{U}_{2/1}$ elements. Besides, construction of the symmetric boxes based on the $\mathbf{Z}'_{2/1}$ elements also requires additional use of the $\mathbf{U}_{2/1}$ elements. Thus, $\mathbf{U}_{2/1}$ elements are also interesting as cryptographic primitive. The $\{\mathbf{U}_{2/1}\}$ subset and main DCs of the $\mathbf{U}_{2/1}$ elements are presented in Table 3.4.

In the following section, we will consider the statistic properties of the $\mathbf{F}_{n/m}$ boxes based on different types of CEs. The results show that the best avalanche caused by changing input bits corresponds to the case of the CSPNs based on the $\mathbf{U}_{2/1}$ and $\mathbf{Q}_{2/1}$ elements. Thus, the $\mathbf{U}_{2/1}$ elements are sufficiently useful as primitive for constructing different variants of the DDOs.

3.2 Controlled Elements Suitable to Field Programmable Gate Array (FPGA) Implementation

In the case of FPGA implementation, all types of the CEs $\mathbf{F}_{2/1}$ are implemented using two 4-bit memory cells, each cell implementing a BF in three variables. However, each of such cells can implement an arbitrary BF in four variables. Pairs of such functions can be used to construct the $\mathbf{F}_{2/2}$ CEs that are controlled with

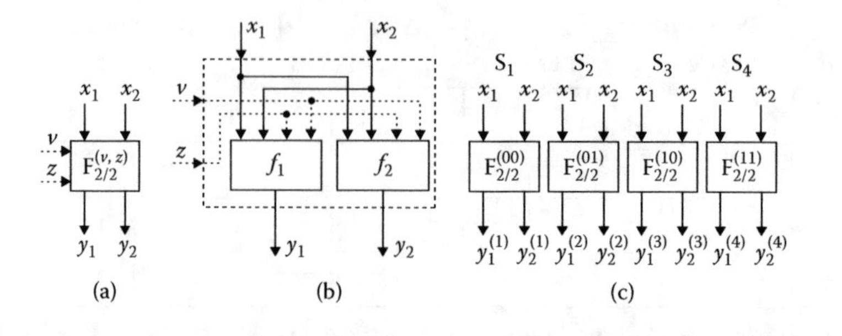

Figure 3.3 Element $F_{2/2}^{(v,z)}$ (*a*) represented as a pair of BFs in four variables (*b*) or as four 2 × 2 substitutions (*c*).

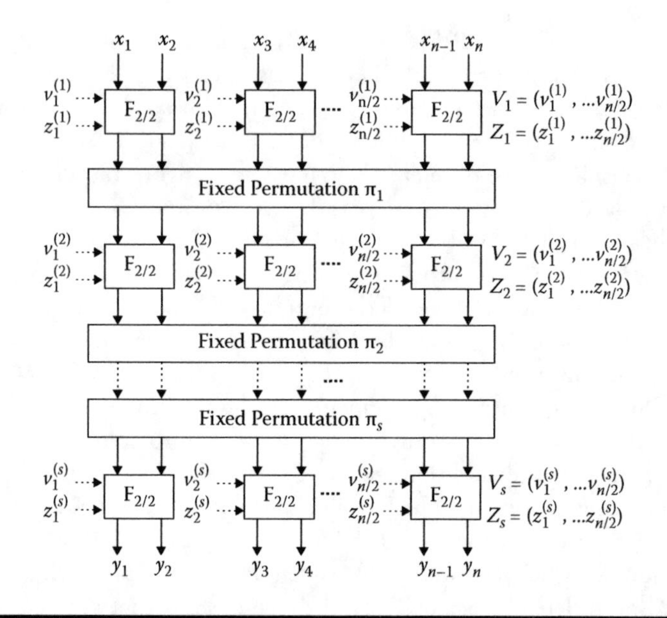

Figure 3.4 General structure of the $\Phi_{n/m}$, boxes.

two-bit controlling vector (v, z) defining selection of one of four possible elementary transformations S_1, S_2, S_3, and S_4 (see Figure 3.3). Such controlled units are properly designed and have higher NL than the $\mathbf{F}_{2/1}$ CEs; therefore, the $\mathbf{F}_{2/2}$ CEs can be used to replace $\mathbf{P}_{2/1}$ elements in the $\mathbf{P}_{n/m}$ boxes or $\mathbf{F}_{2/1}$ elements in the $\mathbf{F}_{n/m}$ boxes, providing construction of more advanced DDO boxes $\mathbf{\Phi}_{n/2m}$. The general structure of such DDO boxes is shown in Figure 3.4. To construct the $\mathbf{\Phi}_{n/2m}$ boxes, we can use the topologies of the $\mathbf{P}_{n/m}$ boxes and three recursive construction mechanisms considered in the preceding text. Two description forms of the $\mathbf{F}_{2/2}$ CEs are connected with the following expressions:

$$y_1 = (v \oplus 1)\ (z \oplus 1)y_1^{(1)} \oplus z(v \oplus 1)y_1^{(2)} \oplus v(z \oplus 1)\ y_1^{(3)} \oplus vzy_1^{(4)} \quad (3.7)$$

$$y_2 = (v \oplus 1)\ (z \oplus 1)y_2^{(1)} \oplus z(v \oplus 1)y_2^{(2)} \oplus v(z \oplus 1)y_2^{(3)} \oplus vzy_2^{(4)} \quad (3.8)$$

Outputs of the $\mathbf{F}_{2/2}$ CEs can also be described as follows:

$$y_1 = vz(y_1^{(1)} \oplus y_1^{(2)} \oplus y_1^{(3)} \oplus y_1^{(4)}) \oplus v(y_1^{(1)} \oplus y_1^{(3)}) \oplus z(y_1^{(1)} \oplus y_1^{(2)}) \oplus y_1^{(1)}; \quad (3.9)$$

$$y_2 = vz(y_2^{(1)} \oplus y_2^{(2)} \oplus y_2^{(3)} \oplus y_2^{(4)}) \oplus v(y_2^{(1)} \oplus y_2^{(3)}) \oplus z(y_2^{(1)} \oplus y_2^{(2)}) \oplus y_2^{(1)}. \quad (3.10)$$

For given values n and s, the FPGA implementation of the $\mathbf{F}_{n/m}$ and $\mathbf{\Phi}_{n/2m}$ boxes requires the use of the same quantity of hardware resources; however, on the whole, the $\mathbf{\Phi}_{n/2m}$ box is a more efficient cryptographic operation.

To select the $\mathbf{F}_{2/2}$ CEs suitable as elementary building blocks in the design of the CSPNs implementing cryptographic transformation operations, it is reasonable to use the following criteria:

1. Each of two outputs of the $\mathbf{F}_{2/2}$ CEs is a nonlinear balanced BF in four variables.
2. Each modification of CEs is a bijective transformation $(x_1, x_2) \rightarrow (y_1, y_2)$ that is an involution.
3. The sum of two outputs, i.e., $f_3 = y_1 \oplus y_2$, is a nonlinear BF in four variables.

In the case of the BFs in three variables, all nonlinear balanced BFs f possess the NL value $\mathrm{NL}(f(x_1, x_2, v)) = 2$. However, in the case of the nonlinear balanced BFs in four variables $f_1(x_1, x_2, v, z)$ we have two different NL values: 2 and 4. Therefore, while classifying the nonlinear elements $\mathbf{F}_{2/2}$, we should take into account different NL values. It is possible to strengthen the NL requirements in the selection criteria and formulate them as follows:

1. Each of two outputs of CEs is a balanced BF in four variables having nonlinearity $\mathrm{NL}(y_1) = \mathrm{NL}(y_2) = 4$.
2. Each modification of CEs is a bijective transformation $(x_1, x_2) \rightarrow (y_1, y_2)$ that is an involution.
3. The sum of two outputs, i.e., $f_3 = y_1 \oplus y_2$, is a BF in four variables having nonlinearity $\mathrm{NL}(f_3) = 4$.

In such a case, we will restrict the number of the $\mathbf{F}_{2/2}$ elements; however, we will restrict simultaneously the set of CEs suitable as cryptographic primitives. For example, the CEs described with BF having NL value 2 can be useful to define deterministic changes in the output caused by inverting one or several bits of the controlling data subblock, such as the $\mathbf{Z}'_{2/1}$ and $\mathrm{U}_{2/1}$ elements. Therefore, the first version of the selection criteria is preferable for the classification of the $\mathbf{F}_{2/2}$ boxes. As in the case of $\mathbf{F}_{2/1}$ boxes, we can use two different approaches:

Table 3.5 Boolean functions in four variables $f(x_1, x_2, x_3, x_4)$ classified on the nonlinearity value NL(f)

NL(f)	6	5	4	3	2	1	0
# BF	896	14336	28000	17920	3840	512	32
# balanced BF	0	0	10920	0	1920	0	30

Table 3.6 Number of the CEs $F_{2/2}$ with different integral nonlinearity value

NL_1-NL_2-NL_3	4-4-4	4-4-2	4-2-2	4-4-0	2-2-2	2-2-0	0-0-0
# CEs	126720	110592	64512	8064	12288	9216	384

Table 3.7 Number of the controlled involutions $F_{2/2}$ with different integral nonlinearity

NL_1-NL_2-NL_3	4-4-4	4-4-2	4-2-2	4-4-0	2-2-2	2-2-0	0-0-0
# CEs	2208	2304	3264	792	384	960	88

1. Consider all possible values of the ordered pairs of the balanced BF in four variables.
2. Consider all possible ordered sets of four elementary substitutions in which we have at least two different substitutions of the 2×2 type.

The number of the balanced BFs in 4 variables is equal to 12,870; therefore, in the first case we should consider more than 10^8 variants. If only nonlinear balanced BFs are considered, then the number of variants is also more then 10^8. Indeed, we have 10,920 functions having NL 4 and 1,920 functions having NL 2 (see Table 3.5).

The exhaustive search of all possible invertible CEs $\mathbf{F}_{2/2}$ satisfying the selection criteria is more efficient in the second case. Indeed, the number of possible sets $\{S_1, S_2, S_3, S_4\}$ is less than 340,000. Using the last approach, the $\mathbf{F}_{2/2}$ elements have been classified relative to their nonlinear properties [44,46]. In their classification, the notion of the integral NL is used, introduced as a set of three numbers, (NL_1, NL_2, NL_3), where $NL_1 = NL(y_1 = f_1(x_1, x_2, v, z))$, $NL_2 = NL(y_2 = f_2(x_1, x_2, v, z))$, and $NL_3 = NL(y_1 \oplus y_2)$. Table 3.6 presents the number of elements corresponding to different subclasses classified according to their integral NL value. Table 3.7 shows the number of elements $\mathbf{F}_{2/2}$ that are involutions. Subclass of the maximum NL CEs $\mathbf{F}_{2/2}$ includes 126,720 elements, among which we have 2,208 involutions.

To get more detailed classification of the $\mathbf{F}_{2/2}$ elements, we should take into account their DCs (see Figure 3.5), which can be estimated with the probability

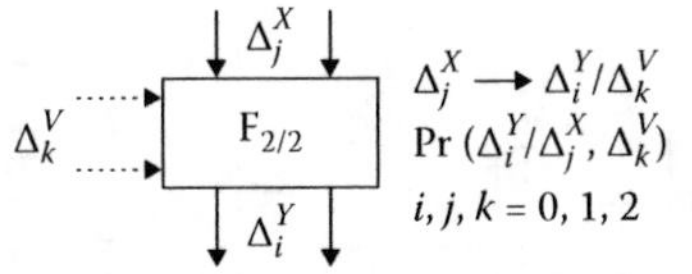

Figure 3.5. Possible differential characteristics of the $F_{2/2}$ elements.

Table 3.8 Characteristic distributions of the probabilities p_{ij0}

i	*j*	*k*	A	B	C	D	E
0	1	0	0	0	0	0	0
1	1	0	¾	⅝	⅞	½	1
2	1	0	¼	⅜	⅛	½	0
0	2	0	0	0	0	0	0
1	2	0	½	¾	¼	1	0
2	2	0	½	¼	¾	0	1

$p_{ijk} = \Pr(\Delta_j^X \rightarrow \Delta_i^Y / \Delta_k^V)$. To estimate all such probabilities corresponding to the same CE using some integral value, the notion of the average entropy has been introduced as follows [46]:

$$\langle \mathrm{H} \rangle = \frac{\left(\sum_{j=0}^{2} \sum_{k=1}^{2} \mathrm{H}_{jk} + \sum_{j=1}^{2} \mathrm{H}_{j0} \right)}{8} \tag{3.11}$$

where the individual sets of the DCs are characterized with the entropy value

$$\mathrm{H}_{jk} = -\sum_{i=0}^{2} p\left(\Delta_i^Y / \Delta_j^X,\ \Delta_k^V\right) \log_3 p\left(\Delta_i^Y / \Delta_j^X,\ \Delta_k^V\right) \tag{3.12}$$

In case $k = 0$, all DCs of the $\mathbf{F}_{2/2}$ elements are attributed to five sets, A, B, C, D, and E, presented in Table 3.8. DCs of the controlled involutions $\mathbf{F}_{2/2}$ having maximum integral NL $(NL_1, NL_2, NL_3) = (4, 4, 4)$ are attributed to four sets, A, B, D, and E. These 2208 elements are divided into 16 sets having different values of the average entropy, which are presented in Table 3.9 (now we have more than 5 sets, as we take into account all of the cases $k = 0, 1, 2$), where the third column shows the 2×2 substitutions implemented with the $\mathbf{F}_{2/2}$ elements of the respective set.

Table 3.9 Subsets of the $F_{2/2}$ involutions having different average entropy values

⟨H⟩	Number of involutions $F_{2/2}$	Variants of the 2×2 substitutions	Types	Examples
0.840	128	e, f, g, h, i, j	(L, R, R, Q)	(e,f,g,h), (f,i,e,j), (h,f,j,e), (j,i,f,e)
0.834	704	a, b, c, d, e, f, g, h, i, j	(L, L, L, L)(Z, L, R, Q)(R, Q, R, L)(L, R, R, L)	(a,d,g,i), (b,i,c,h), (f,i,e,h), (j,i,f,d)
0.815	128	a, b, c, d, g, h, i, j	(L, L, L, L)	(a,b,j,g), (b,d,h,i), (h,c,i,a), (j,g,c,d)
0.813	192	a, d, e, f, g, h, i, j	(L, L, R, R)(L, R, R, L)	(e,e,g,j), (f,h,f,h), (h,h,e,e), (j,f,g,f)
0.812	256	b, c, e, f, g, h, i, j	(Z, L, R, Q)	(b,e,g,h), (c,i,e,j), (e,b,h,j), (i,c,j,e)
0.791	128	e, f, g, h, i, j	(R, R, Q, Q)	(e,g,j,h), (f,h,i,g), (g,h,h,e), (j,i,i,f)
0.788	128	b, c, e, f, g, h, i, j	(L, L, R, R)	(b,g,h,e), (c,h,j,f), (h,e,c,g), (j,f,b,i)
0.786	64	e, f, g, h, i, j	(R, R, R, R)	(e,g,h,f), (g,f,f,i), (i,e,e,j), (j,f,e,h)
0.774	32	g, h, i, j	(L, Q, Q, L)	(g,g,h,i), (h,i,j,j), (i,j,h,j), (j,i,j,h)
0.731	32	g, h, i, j	(L, Q, Q, L)	(g,g,h,h), (h,g,i,j), (i,g,i,g), (j,g,i,h)
0.719	96	a, b, c, d, e, f, g, h, i, j	(L, L, L, L)(R, R, R, R)	(a,g,g,d), (b,h,i,c), (h,c,b,h), (j,f,e,g)
0.710	16	g, h, i, j	(Q, Q, Q, Q)	(f,h,e,i), (h,j,j,i), (i,g,j,i), (j,i,i,g)
0.695	192	a, b, c, d, e, f, g, h, i, j	(L, L, L, L)(L, Z, Z, L)	(a,b,e,e), (d,f,c,e), (g,d,c,g), (j,c,d,j)
0.641	32	e, f, g, h, i, j	(R, R, R, R)	(e,h,i,e), (f,g,g,f), (h,e,e,i), (j,f,f,g)
0.631	64	a, b, c, d, e, f	(Z, Z, Z, Z)	(a,e,e,b), (c,e,f,a), (e,c,a,f), (f,d,c,f)
0.513	16	g, h, i, j	(Q, Q, Q, Q)	(g,h,h,g), (h,g,j,i), (i,g,g,i), (j,i,h,g)

While fixing the controlling bit z in the box $\mathbf{F}_{2/2}$, we transform the $\mathbf{F}_{2/2}$ CE into the $\mathbf{F}_{2/1}$ element, implementing a pair of the 2 × 2 substitutions $\mathbf{F}_{2/2}^{(0,z)}$ and $\mathbf{F}_{2/2}^{(1,z)}$. For two different values $z = 0$ and $z = 1$, we have, in general, two different variants of the $\mathbf{F}_{2/1}$ elements. If we fix the controlling bit v, then we get some other two different elements $\mathbf{F}_{2/1}$ (for $v = 0$ and $v = 1$), each of which implements the pair of the 2 × 2 substitutions $\mathbf{F}_{2/2}^{(v,0)}$ and $\mathbf{F}_{2/2}^{(v,1)}$. Thus, we can represent the $\mathbf{F}_{2/2}$ CE as a primitive implementing two modifications, each of which is an elementary controlled operation ($\mathbf{F}_{2/1}$), and we have two different variants of such representation. The controlled modifications $\mathbf{F}_{2/2}^{(0,z)}$, $\mathbf{F}_{2/2}^{(0,z)}$ with controlling bit z, and $\mathbf{F}_{2/2}^{(v,0)}$, $\mathbf{F}_{2/2}^{(v,1)}$ with controlling bit v, can be attributed to the $\mathbf{R}_{2/1}$, $\mathbf{Q}_{2/1}$, $\mathbf{Z}_{2/1}$ types, or some other types. Therefore, we can give some formal characterization of the $\mathbf{F}_{2/2}$ elements as sets of four letters denoting the types of the indicated controlled modifications, for example, (Q, Q, Q, Q), (R, R, R, R), and (R, R, Q, Q). In general, some of the $\mathbf{F}_{2/1}$ elements as elementary controlled modifications of the $\mathbf{F}_{2/2}$ elements can be out of the NL criteria introduced to select the $\mathbf{F}_{2/1}$ elements as primitives to designing controlled operations. Therefore, considering the $\mathbf{F}_{2/2}$ CE we should take into account different other types of the $\mathbf{F}_{2/1}$ elements, including linear elements $\mathbf{L}_{2/1}$, i.e., elements having one linear outut. Indeed, there exist elements $\mathbf{F}_{2/2}$ with values $\langle H \rangle = 0.834$ and $(NL_1, NL_2, NL_3) = (4, 4, 4)$, which can be represented as (L, L, L, L). More often we meet cases when only some of four $\mathbf{F}_{2/1}$-modifications are nonlinear, for example, (L, R, R, Q), (Z, L, R, Q), (L, L, R, R), and (L, Q, Q, L).

3.3 Symmetric Topologies

General structure of the CSPN based on the $\mathbf{F}_{2/1}$ elements can be represented as some superposition of active layers and fixed permutations, as in the case of the CP boxes. Thus, by the term *active layer*, we understand a cascade of the $\mathbf{F}_{2/1}$ elements that operate on data bits in parallel. Usually, the active layers containing $n = 2^z$ controlled elements, where z is a natural number, are considered in this book. They are denoted as $\mathbf{L}$, $\mathbf{L}_n$, or $\mathbf{L}_{(z)}$. General structure of some $\mathbf{F}_{n/m}$ boxes and their inverses can be easily derived from the general structure of the $\mathbf{P}_{n/m}$ and $\mathbf{P}_{n/m}^{-1}$ boxes. The notion of the order introduced for DDP boxes can be extended in the following way:

Definition 3.1. *A DDO box constructed using the* $\mathbf{F}_{2/1}$ *elements as standard building blocks is called a box of the order h if it has the same topology of the connections between active layers as some CP box of the order h.*

Thus, three recursive construction mechanisms considered in Chapter 2 (see Figures 2.2*a*,*b* and 2.3) and other topologies of the CP boxes can be used to construct the $\mathbf{F}_{n/m}$ boxes of different orders. Topologies having mirror symmetry represent significant interest for designing iterated block ciphers. While considering $\mathbf{F}_{n/m}$ boxes, we should take into account the fact that, in general, the $\mathbf{F}_{2/1}$ elements are not involutions. Therefore, in symmetric topologies we should use at least two different variants of the $\mathbf{F}_{2/1}$ elements (in some cases, these CEs can correspond to different types) if they are not involutions. We define the symmetric boxes $\mathbf{F}_{n/m}$ as follows:

Definition 3.2. *A topology of the* $\mathbf{F}_{n/m}$ *box is called symmetric if for all* $j = 1, ..., s-1$ *the following relations hold:* $\mathbf{L}_j = (\mathbf{L}_{s-j+1})^{-1}$ (*or* $\mathbf{L}_j = \mathbf{L}_{s-j+1}$, *if* $\mathbf{L}_j$ *is involution*) and $\pi_j = (\pi_{s-j})^{-1}$, *where j denotes sequential number of the active layers in the s-layer* $\mathbf{F}_{n/m}$ *box.*

Definition 3.3. *An s-layer* $\mathbf{F}^{(V)}_{n/m}$ *box having symmetric topology is called symmetric if for all* $j = 1, ..., s-1$ *the following relations hold:* $V_j = V_{s-j+1}$, *where j denotes sequential number of the active layers in the s-layer* $\mathbf{F}_{n/m}$ *box.*

If the number of active layers is odd, then in a symmetric box the $\mathbf{L}_{[s/2]+1}$ layer is an involution, i.e., it is composed of CEs that are involutions. If some $\mathbf{F}_{n/m}$ box having symmetric structure contains $\mathbf{F}_{2/1}$ elements that are not involutions, then it also contains at least the $\mathbf{F}^{-1}_{2/1}$ elements. If, in such $\mathbf{F}_{n/m}$ box we have odd number of active layers, then it contains at least three different variants of the CEs: $\mathbf{F}_{2/1}$, $\mathbf{F}^{-1}_{2/1}$, and elementary controlled involutions. In Chapter 2 we have considered block ciphers Cobra-H64 and DDP-64 using the $\mathbf{P}_{32/96}$ box having symmetric structure. Replacing in this box all switching elements by controlled involutions $\mathbf{Q}_{2/1}$, $\mathbf{R}_{2/1}$, $\mathbf{Z}_{2/1}$, or $\mathbf{U}_{2/1}$ of the same kind, we get different types of the symmetric uniform (homogeneous) $\mathbf{F}_{32/96}$ boxes of the second order. In a general case, using the boxes $\mathbf{F}_{2/1}$ and $\mathbf{F}^{-1}_{2/1}$, we get symmetric $\mathbf{F}_{32/96}$ boxes with topology shown in Figure 3.6.

Replacing in the $\mathbf{F}_{8/12}$ and $\mathbf{F}^{-1}_{8/12}$ boxes $\mathbf{F}_{2/1}$ and $\mathbf{F}^{-1}_{2/1}$ elements by mutually inverse elements $\mathbf{F}_{2/2}$ and $\mathbf{F}^{-1}_{2/2}$, respectively, we get uniform $\mathbf{\Phi}_{8/12}$ and $\mathbf{\Phi}^{-1}_{8/12}$ boxes of the first order. The $\mathbf{\Phi}_{8/12}$ and $\mathbf{\Phi}^{-1}_{8/12}$ boxes can be used to construct the second order box $\mathbf{\Phi}_{32/192}$ (Figure 3.7*a*) and the first order box $\mathbf{\Phi}_{64/384}$ (Figure 3.7*c*). Using elementary controlled involutions $\mathbf{F}_{2/2}$, we can get uniform symmetric boxes $\mathbf{\Phi}_{32/192}$ and $\mathbf{\Phi}_{64/384}$ with the topologies presented in Figure 3.7. (In this case, the $\mathbf{\Phi}_{8/12}$ and $\mathbf{\Phi}^{-1}_{8/12}$ boxes are constructed using the same elements $\mathbf{F}_{2/2}$.)

The $\mathbf{\Phi}_{32/96}$ and $\mathbf{\Phi}_{64/384}$ boxes can be represented as the following superpositions:

$$\mathbf{\Phi}_{32/192} = \mathbf{\Phi}_{32/96} \circ \mathbf{I}_1 \circ \mathbf{\Phi}^{-1}_{32/96} \tag{3.13}$$

$$\mathbf{\Phi}_{64/384} = \mathbf{\Phi}_{64/192} \circ \mathbf{I}_2 \circ \mathbf{\Phi}^{-1}_{64/192} \tag{3.14}$$

where $\mathbf{I}_1$ and $\mathbf{I}_2$ are the following permutational involutions:

$\mathbf{I}_1$: (1)(2,9)(3,17)(4,25)(5)(6,13)(7,21)(8,29)(10)(11,18)(12,26)
(14)(15,22)(16,30)(19)(20,27)(23)(24,31)(28)(32);

$\mathbf{I}_2$: (1)(2,9)(3,17)(4,25)(5,33)(6,41)(7,49)(8,57)(10)(11,18)(12,26)(13,34)(14,42)
(15,50)(16,58)(19)(20,27)(21,35)(22,43)(23,51)(24,59)(28)(29,36)(30,44)(31,52)
(32,60)(37)(38,45)(39,53)(40,61)(46)(47,54)(48,62)(55)(56,63)(64).

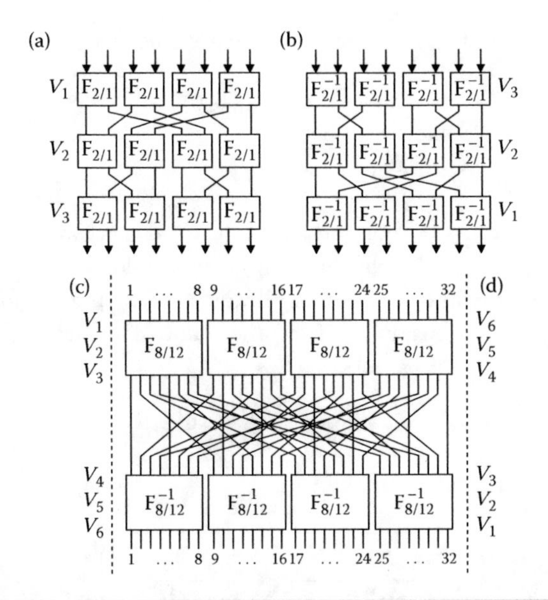

Figure 3.6 Topology of the boxes $F_{8/12}$ (*a*), $F_{8/12}^{-1}$ (*b*), $F_{32/96}$ (*c*), and $F_{32/96}^{-1}$ (*d*).

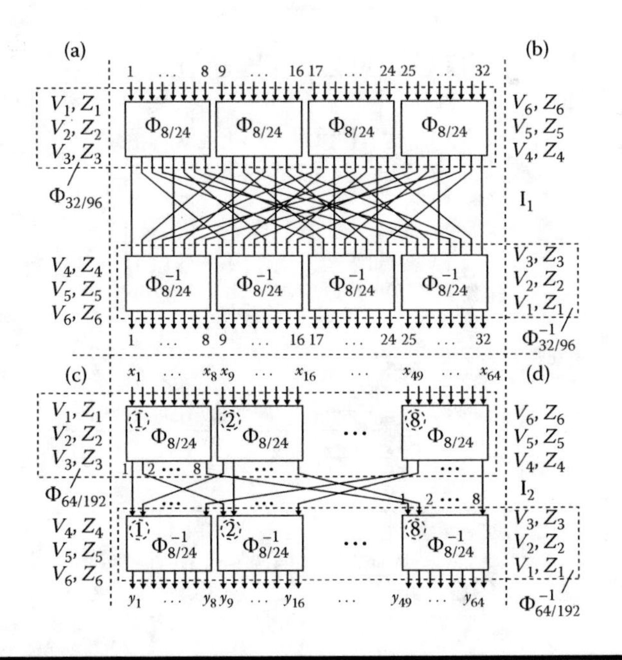

Figure 3.7 Structure of the $\Phi_{32/96}$ (*a*), $\Phi_{32/96}^{-1}$ (*b*), $\Phi_{64/384}$ (*c*), and $\Phi_{64/384}^{-1}$ (*d*) boxes.

Similarly, the $\mathbf{F}_{32/96}$ and $\mathbf{F}_{64/192}$ boxes can be represented as

$$\mathbf{F}_{32/96} = \mathbf{F}_{32/48} \circ \mathbf{I}_1 \circ \mathbf{F}^{-1}_{32/48} \tag{3.15}$$

$$\mathbf{F}_{64/192} = \mathbf{F}_{64/96} \circ \mathbf{I}_2 \circ \mathbf{F}^{-1}_{64/96} \tag{3.16}$$

At each step of the recursive construction of the third type, we have the maximum order ($h = n$) $\mathbf{F}_{n/m}$ (or $\mathbf{\Phi}_{n/m'}$) box having symmetric topology.

Let us consider the next design. For all values $n = 2^z$ and $h = n/2$, it is possible to construct symmetric box $\mathbf{F}_{2n/4m}$ using the following generalized topology (see Figure 3.8*a*):

$$(\mathbf{F}_{n/m} \| \mathbf{F}_{n/m}) \circ \mathbf{I}_3 \circ (\mathbf{F}^{-1}_{n/m} \| \mathbf{F}^{-1}_{n/m}) \tag{3.17}$$

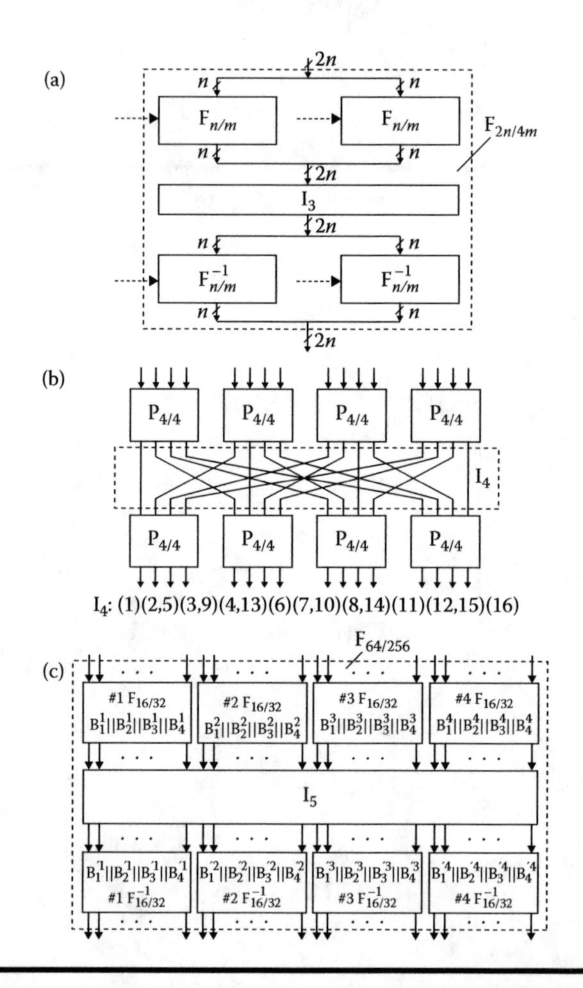

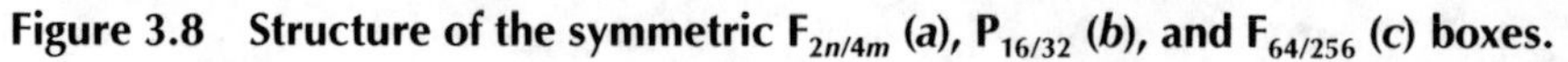

Figure 3.8 Structure of the symmetric $F_{2n/4m}$ (*a*), $P_{16/32}$ (*b*), and $F_{64/256}$ (*c*) boxes.

where $\mathbf{F}_{n/m}||\mathbf{F}_{n/m}$ denotes concatenation of two boxes $\mathbf{F}_{n/m}$ in a single cascade. The $\mathbf{F}_{n/m}$ and $\mathbf{F}_{n/m}^{-1}$ are mutually inverse boxes of the first order, and permutational involution $\mathbf{I}_3$ is described as follows:

$$\mathbf{I}_3: (1) \ldots (2i-1) \ldots (n/2-1)(2, n/2+1)\ldots$$

$$\ldots (2i, 2i+n/2-1) \ldots (n/2, n-1)(n/2+2) \ldots (2j+n/2) \ldots (n),$$

where $i = 1, 2, \ldots, n/4$, and $j = 1, 2, \ldots, n/4$. For example, in the case $n = 32$, we have

$$\mathbf{I}_3': (1) \ldots (2i-1) \ldots (15)(2,17) \ldots (2i, 2i+15) \ldots (16,31)(18) \ldots (2j+16) \ldots (32),$$

where $i = 1, 2, \ldots, 8$ and $j = 1, 2, \ldots, 8$. Because the $\mathbf{I}_3'$ permutation is out of the set of the fixed permutations corresponding to the first, second, and third recursive construction mechanisms, the considered $\mathbf{F}_{2n/4m}$ topology is different from the topologies derived with the indicated mechanisms.

In general, for $h = n/4, n/16, n/64 \ldots$ (in this case we have even number of active layers in the recursive topology of the network), it is possible to construct the $\mathbf{F}_{n/m}$ boxes having mirror-symmetry topology and satisfying condition $s = \log_2 nh$ (this is the minimum number of active layers in some networks with the size of input n and order h). Some of such symmetric boxes are presented in Table 3.10, in which we can see the following diagonal rows:

Table 3.10 The CP boxes with symmetric structure

n \ h	1	2	4	8	16	32	64	128	256
2	—	$\mathbf{P}_{2/1}$	—	—	—	—	—	—	—
4	$\mathbf{P}_{4/4}$	—	$\mathbf{P}_{4/6}$	—	—	—	—	—	—
8	—	$\mathbf{P}_{8/16}$	—	$\mathbf{P}_{8/20}$	—	—	—	—	—
16	$\mathbf{P}_{16/32}$	—	$\mathbf{P}_{16/48}$	—	$\mathbf{P}_{16/56}$	—	—	—	—
32	—	$\mathbf{P}_{32/96}$	—	$\mathbf{P}_{32/128}$	—	$\mathbf{P}_{32/144}$	—	—	—
64	$\mathbf{P}_{64/192}$	—	$\mathbf{P}_{64/256}$	—	$\mathbf{P}_{64/320}$	—	$\mathbf{P}_{64/352}$	—	—
128	—	$\mathbf{P}_{128/512}$	—	$\mathbf{P}_{128/640}$	—	$\mathbf{P}_{128/768}$	—	$\mathbf{P}_{128/832}$	—
256	$\mathbf{P}_{256/1024}$	—	$\mathbf{P}_{256/1280}$	—	$\mathbf{P}_{256/1536}$	—	$\mathbf{P}_{256/1792}$	—	$\mathbf{P}_{256/2048}$

a. $\mathbf{P}_{2/1} \rightarrow \mathbf{P}_{4/6} \rightarrow \mathbf{P}_{8/20} \rightarrow \mathbf{P}_{16/56} \rightarrow \mathbf{P}_{32/144} \rightarrow \mathbf{P}_{64/352} \rightarrow \mathbf{P}_{128/832} \rightarrow \ldots$ (this row corresponds to the maximum order boxes; it is constructed using the recursive mechanism of the third type; at the first step, the maximum order box $\mathbf{P}_{2/1}$ is used)

b. $\mathbf{P}_{4/4} \rightarrow \mathbf{P}_{8/16} \rightarrow \mathbf{P}_{16/48} \rightarrow \mathbf{P}_{32/128} \rightarrow \mathbf{P}_{64/320} \rightarrow \mathbf{P}_{128/768} \rightarrow \ldots$ (this row corresponds to the order value $h = n/4$; it is constructed using the recursive mechanism of the third type; at the first step, the first order box $\mathbf{P}_{4/4}$ is used)

c. $\mathbf{P}_{16/32} \rightarrow \mathbf{P}_{32/96} \rightarrow \mathbf{P}_{64/256} \rightarrow \mathbf{P}_{128/640} \rightarrow \ldots$ (this row corresponds to the order value $h = n/16$; it is constructed using the recursive mechanism of the third type; at the first step, the first order box $\mathbf{P}_{16/32}$ is used, which is shown in Figure 3.8*b*)

d. $\mathbf{P}_{64/192} \rightarrow \mathbf{P}_{128/512} \rightarrow \ldots$ (this row corresponds to the order value $h = n/64$; it is constructed using the recursive mechanism of the third type; at the first step, the first order box $\mathbf{P}_{64/192}$ is used)

The first two rows are attributed to the topologies corresponding to the "standard" recursive mechanism, i.e., to the mechanisms of the first, second, and third types. The last two rows are attributed to other construction mechanisms, because at the first step of the construction, initial boxes have topologies that are different from "standard" ones. In general, for given values h and n, different symmetric topologies are possible, which provide minimum number of active layers. The symmetric topologies of the CP boxes presented in Table 3.10 can be used to design different types of symmetric $\mathbf{F}_{n/m}$ and $\mathbf{\Phi}_{n/m}$ boxes.

Besides the topologies presented in Table 3.10, there are also other symmetric constructions providing minimum number of active layers for a given number n. For example, the fourth-order $\mathbf{F}_{64/256}$ box (see Figure 3.8*c*) with symmetric topology can be constructed also using the following structure:

$$\mathbf{F}_{64/256} = (\mathbf{F}_{16/32} \| \mathbf{F}_{16/32} \| \mathbf{F}_{16/32} \| \mathbf{F}_{16/32}) \circ \mathbf{I}_5 \circ (\mathbf{F}^{-1}_{16/32} \| \mathbf{F}^{-1}_{16/32} \| \mathbf{F}^{-1}_{16/32} \| \mathbf{F}^{-1}_{16/32}), \quad (3.18)$$

where permutational involution $\mathbf{I}_5$ is described as follows:

$\mathbf{I}_5$: (1)(2, 17)(3, 33)(4, 49)(5)(6, 21)(7, 37)(8, 53)(9)(10, 25)(11, 41)(12, 57)(13)
(14, 29)(15, 45)(16, 61)(18)(19, 34)(20, 50)(22)(23, 38)(24, 54)(26)(27, 42)
(28, 58)(30)
(31, 46)(32, 62)(35)(36, 51)(39)(40, 55)(43)(44, 59)(47)(48, 63)(52)(56)(60)(64).

The $\mathbf{I}_5$ permutation can be interpreted as concatenation of four permutational involutions $\mathbf{I}_4$ used to connect the upper and lower cascades of the $\mathbf{P}_{4/4}$ boxes in the first-order $\mathbf{P}_{16/32}$ box (see Figure 3.8*b*). Indeed, the $\mathbf{I}_4$ permutation can be described as

$$\mathbf{I}'_4 : (a)(b,e)(c,i)(d,m)(f)(g,j)(h,n)(k)(l,o)(p).$$

For arbitrary $z \in \{1, 2, 3, 4\}$, the $\mathbf{I}'_4$ permutation links four boxes $\mathbf{B}_z^1$, $\mathbf{B}_z^2$, $\mathbf{B}_z^3$, and $\mathbf{B}_z^4$ with four boxes $\mathbf{B}_z'^1$, $\mathbf{B}_z'^2$, $\mathbf{B}_z'^3$, and $\mathbf{B}_z'^4$. For example, in the case $z = 1$, we have a = 1, b = 2, c = 3, d = 4, e = 17, f = 18, g = 19, h = 20, i = 33, j = 34, k = 35, l = 36, m = 49, n = 50, o = 51, p = 52.

Two lower active layers of the *i*th box $\mathbf{F}_{16/32}$ compose a cascade $\mathbf{B}_1^i||\mathbf{B}_2^i||\mathbf{B}_3^i||\mathbf{B}_4^i$ of four boxes $\mathbf{F}_{4/4}$, and two upper active layers of the *j*th box $\mathbf{F}_{16/32}^{-1}$ compose a cascade $\mathbf{B}_1'^j||\mathbf{B}_2'^j||\mathbf{B}_3'^j||\mathbf{B}_4'^j||$ of four boxes $\mathbf{F}_{4/4}$. For each value $z \in \{1, 2, 3, 4\}$, the boxes $\mathbf{B}_z^i$, $i = 1, 2, 3, 4$, are linked with the boxes $\mathbf{B}_z'^j$ in accordance with the rule "the *j*th output of the box $\mathbf{B}_z^i$ is connected with the *i*th input of the box $\mathbf{B}_z'^j$, $j = 1, 2, 3, 4$." Thus, each of the $\mathbf{F}_{16/32}$ boxes has four connections with each of the $\mathbf{F}_{16/32}^{-1}$ boxes.

It is easy to demonstrate that symmetric box $\mathbf{F}_{64/256}$ has order 4. Indeed, let us consider the CP box $\mathbf{P}_{64/256}$ having the same topology as the described box $\mathbf{F}_{64/256}$. Arbitrary four input bits of the *i*th box $\mathbf{P}_{16/32}$ can be fed to input of four boxes $\mathbf{B}_1^i||\mathbf{B}_2^i||\mathbf{B}_3^i||\mathbf{B}_4^i$ (in $\mathbf{P}_{64/256}$, the **B** box represents the $\mathbf{P}_{4/4}$ box); therefore, each of these four bits can be fed to input of arbitrary box $\mathbf{P}_{16/32}^{-1}$ in the cascade $\mathbf{P}_{16/32}^{-1}||\mathbf{P}_{16/32}^{-1}||\mathbf{P}_{16/32}^{-1}||\mathbf{P}_{16/32}^{-1}$ independently of other bits. If several of these bits are fed to the same box $\mathbf{P}_{16/32}^{-1}$, then all of them are fed to different boxes $\mathbf{B}'$. For example, four bits can be fed to the same box $\mathbf{P}_{16/32}^{-1}$. Because they are fed to different boxes $\mathbf{B}'$, they can be moved to arbitrary four output digits of the $\mathbf{P}_{16/32}^{-1}$ box.

3.4 Properties of the CSPNs Based on Elements $\mathbf{F}_{2/1}$ and $\mathbf{F}_{2/2}$

We should remember that the $\mathbf{F}_{n/m}$ boxes are nonlinear cryptographic primitive only if the controlling vector *V* depends on some data subblock. In other words, the $\mathbf{F}_{n/m}$ boxes are to be used to perform DDOs. In different iterative cryptoschemes, the data subblock is divided into two subblocks of the same length *n*: the controlling one (*L*) and the transformed one (*X*). Thus, one data subblock is used to form the controlling vector *V*, while the other one is transformed with some DDO box. Usually, the controlling subblock is fed to the input of an extension box **E**, implemented in hardware as simple wiring. The output of the **E** box is used as vector *V*. The bit length of the controlling vector is $m = sn/2$ for the $\mathbf{F}_{n/m}$ and $\mathbf{P}_{n/m}$ boxes, and $m' = sn$ for the $\mathbf{\Phi}_{n/m'}$ boxes (*s* denotes number of active layers). This means that each bit of the controlling data subblock determines $z = s/2$ and $z = s$ bits of the *V* vector controlling the $\mathbf{F}_{n/m}$ and $\mathbf{\Phi}_{n/m'}$ boxes, correspondingly. In the case of the $\mathbf{F}_{32/96}$, $\mathbf{F}_{64/192}$, $\mathbf{P}_{32/96}$, and $\mathbf{P}_{64/192}$ boxes, we have $z = 3$. For the $\mathbf{\Phi}_{32/192}$ and $\mathbf{\Phi}_{64/384}$ boxes we have $z = 6$. While designing **E** boxes, to provide each bit of the controlling data subblock influences uniformly to the output of the DDO box, we can use Criteria 2.1 and 2.2 (see Chapter 2). In the case of the $\mathbf{F}_{32/96}$, $\mathbf{F}_{64/192}$, $\mathbf{P}_{32/96}$, and $\mathbf{P}_{64/192}$ boxes, these criteria can be interpreted as follows:

Criterion 2.1′. *Let $X = (x_1, ..., x_n)$ be the input vector of the $\mathbf{F}^{(V)}_{32/96}$ and $\mathbf{F}^{(V)}_{64/192}$ boxes. Then, for all L and i, the bit x_i should pass through six CEs controlled with six different bits of $L = (l_l, ..., l_n)$.*

Criterion 2.2′. *For all i, the bit l_i should define exactly three bits of $V = (l_l, ..., l_m)$.*

In the case of the $\mathbf{\Phi}_{n/m'}$ boxes, the criteria get the following form:

Criterion 2.1″. *For all L and i, the bit x_i should pass through six CEs controlled with twelve different bits of $L = (l_l, ..., l_n)$.*

Criterion 2.2″. *For all i, the bit l_i should define exactly six bits of $V = (l_l, ..., l_m)$.*
For example, in the case of the six-layer boxes $\mathbf{F}_{n/m}$, the extension box $\mathbf{E}$ can be defined as follows:

$$V_1 = L_l;\ V_2 = L_l^{<<<6};\ V_3 = L_l^{<<<12};\ V_4 = L_h^{<<<12};\ V_5 = L_h^{<<<6};\ V_6 = L_h, \quad (3.19)$$

where $V = (V_1, V_2, \ldots, V_6)$; $L = (L_l, L_h) \in \{0, 1\}^n$; $L_l = (l_1, ..., l_{n/2})$; $L_h = (l_{n/2+1}, ..., l_n)$; and $Y = X^{<<<k}$ denotes left rotation of the n-bit word X by k bits, for which we have $y_i = x_{i+k}$ for $1 \le i \le n - k$ and $y_i = x_{i+k-n}$ for $n - k + 1 \le i \le n$. A feature of this extension box is symmetric distribution of the controlling bits corresponding to different halves of the data subblock L.

If Criteria 2.1 and 2.2 are satisfied, then each bit of the controlling data subblock influences s input bits of the $\mathbf{F}_{n/m}$ box and $2s$ input bits of the $\mathbf{\Phi}_{n/m'}$ box. Therefore, minor changes in the subblock L cause significant avalanche. For different variants of the CEs, the contribution to the avalanche of the changes at the controlling input is significantly larger than the contribution of the changes at the input of the DDO box. The CP boxes represent some special case: changes at the input of the $\mathbf{P}_{n/m}$ boxes cause no avalanche. Only changes at the controlling input of the $\mathbf{P}_{n/m}$ boxes contribute to the avalanche effect.

3.4.1 Nonlinearity and Avalanche Properties of the DDP-Like Boxes

Each output of the first order boxes $\mathbf{F}_{n/m}$ is described by a BF in $\mu = 2n - 1$ variables, which has nonlinearity (NL) as follows [45]:

$$\mathrm{NL}(y_i) = 2^{2n-2-\log_2 n}(n-1) \quad (3.20)$$

Maximum NL value of the balanced BF in μ variables is restricted by the estimation [62]:

$$\mathrm{NL}(f_{\mathrm{bal}}) < 2^{\mu-1} - 2^{\frac{\mu}{2}-1} \tag{3.21}$$

As the n value increases, the $\mathrm{NL}(y_i)$ value approximates to maximum possible NL of the balanced BF in the same number of variables:

$$\frac{\mathrm{NL}(y_i)}{\mathrm{NL}_{\max}(f_{\mathrm{bal}})} > \frac{1-\frac{1}{n}}{1-2^{-n+\frac{1}{2}}} \tag{3.22}$$

This shows that even DDO boxes of the first order have sufficiently high NL.

Avalanche of the $\mathbf{F}_{n/m}$ boxes depends significantly on the type of the CEs used as elementary building blocks. Figures 3.9 and 3.10 present the comparison of the probability distribution $p(t) = \Pr(\Delta_t^Y/\Delta_1^X, \Delta_0^V)$ for the $\mathbf{R}_{32/96}$, $\mathbf{Q}_{32/96}$, $\mathbf{P}_{32/96}$, $\mathbf{U}_{32/96}$, and $\mathbf{Z}_{32/96}$ boxes (the $\mathbf{F}_{64/192}$ boxes have the distribution like respective $\mathbf{F}_{32/96}$ boxes [43]). These results correspond to the experiment in which the uniformly distributed random values X and V have been used. One can see that alteration of one input bit leads to deterministic alteration of one output bit for $\mathbf{P}_{32/96}$, whereas for all other boxes many bits are inverted with high probability. The $\mathbf{U}_{32/96}$ and $\mathbf{Q}_{32/96}$ boxes possess about the same distribution of $p(t)$. It is interesting that the uniform boxes $\mathbf{U}_{32/96}$ and $\mathbf{Z}_{32/96}$ built up from mutually inverse CEs $\mathbf{U}_{2/1}$ and $\mathbf{Z}'_{2/1}$, respectively, have significantly different distributions although both of these elements possess the same integral DCs $\Pr(\Delta_1^X \rightarrow \Delta_1^Y/\Delta_0^V) = ½$ and $\Pr(\Delta_1^X \rightarrow \Delta_2^Y/\Delta_0^V) = ½$. In the case $\mathbf{U}_{32/96}$, we have a "smooth" distribution $p(t)$, whereas in the case $\mathbf{Z}_{32/96}$ the distribution is "interrupted." Such differences can be easily explained considering the pairs of the 2 × 2 substitution modifications implemented by the $\mathbf{U}_{2/1}$ and

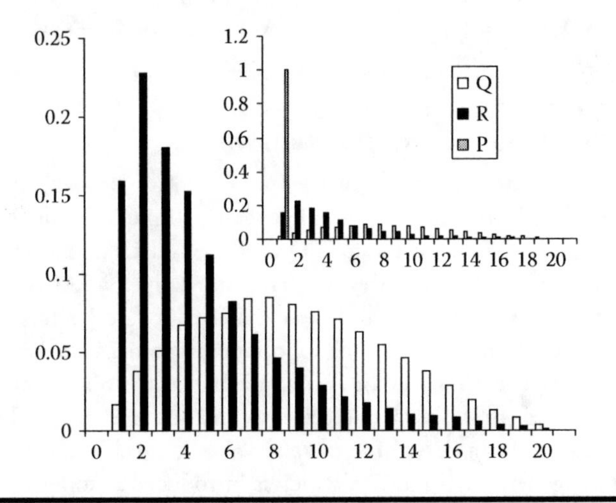

Figure 3.9 Distribution of the probability $p(t) = \Pr(\Delta_t^Y/\Delta_1^X, \Delta_0^V)$ for uniform boxes $F_{32/96}$ built up using controlled elements $Q_{2/1}$, $R_{2/1}$, and $P_{2/1}$.

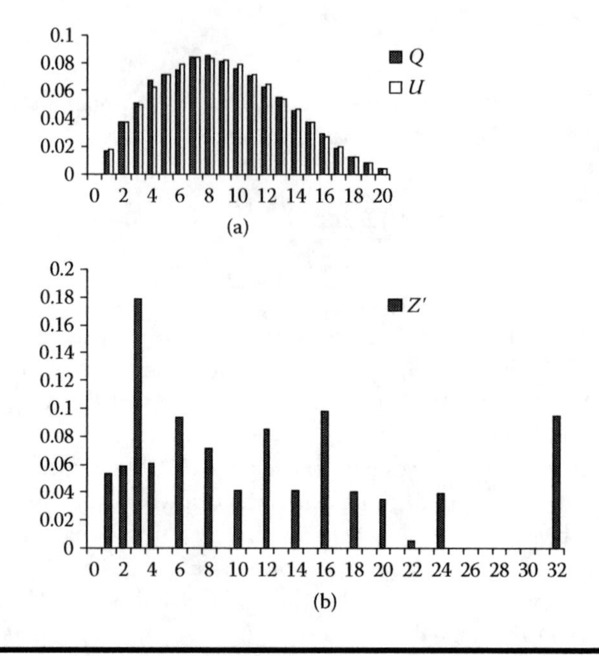

Figure 3.10 Distribution of the probability $p(t) = \Pr(\Delta_t^Y/\Delta_1^X, \Delta_0^V)$ for uniform boxes $Q_{32/96}$, $U_{32/96}$ (*a*), and $Z'_{32/96}$ (*b*).

$\mathbf{Z}'_{2/1}$ elements. Indeed, we can see that $\mathbf{Z}'_{2/1}$ and $\mathbf{U}_{2/1}$ CEs have different DCs with fixed positions of active bits, i.e., for $\mathbf{Z}'_{2/1}$ we have $\Pr(\Delta_{1|1}^X \to \Delta_1^Y/\Delta_0^V) = 0$, $\Pr(\Delta_{1|2}^X \to \Delta_1^Y/\Delta_0^V) = 1$, $\Pr(\Delta_{1|1}^X \to \Delta_2^Y/\Delta_0^V) = 1$, and $\Pr(\Delta_{1|2}^X \to \Delta_2^Y/\Delta_0^V) = 0$; however, for $\mathbf{U}_{2/1}$ we have $\Pr(\Delta_{1|1}^X \to \Delta_1^Y/\Delta_0^V) = ½$, $\Pr(\Delta_{1|2}^X \to \Delta_1^Y/\Delta_0^V) = ½$, $\Pr(\Delta_{1|1}^X \to \Delta_2^Y/\Delta_0^V) = ½$, and $\Pr(\Delta_{1|2}^X \to \Delta_2^Y/\Delta_0^V) = ½$.

3.4.2 Using the Generating Functions

For the $\mathbf{R}_{n/m}$, $\mathbf{Q}_{n/m}$, and $\mathbf{U}_{n/m}$ boxes of the first order, the probability distribution $p(t)$ can be calculated using the method of generating functions and values p_{110} and p_{210} (see Tables 3.1 and 3.4) for the CEs $\mathbf{Q}_{2/1}$, $\mathbf{R}_{2/1}$, and $\mathbf{U}_{2/1}$. Indeed, at the input of the first active layer of such boxes, we have one active bit. At the input of the *i*th layer, we have, with some probabilities, from 1 to 2^{i-1} active bits (i.e., inverted bits), and no pair of the active bits is fed to input of the same CE; therefore, in each layer the events of generating one or two active bits at the output of each CE are independent; besides, they are characterized with the same probabilities p_{110} and p_{210}. In such cases, we can determine the $p(t)$ value as the k_t coefficient in the $k_t x^t$ summand of the generating function, which is introduced as follows:

$$\varphi_1(x) = p_{110}\,x + p_{210}\,x^2;\ \varphi_2(x) = p_{110}\varphi_1(x) + p_{210}\varphi^2{}_1(x);\ \ldots \tag{3.23}$$

$$\varphi_s(x) = p_{110}\varphi_{s-1}(x) + p_{210}\varphi^2{}_{s-1}(x) \tag{3.24}$$

where $\varphi_s(x)$ is the generating function describing the s-layer box $\mathbf{F}_{n/m}$, and x is the formal variable. Thus, $\varphi_1(x)$ is the generating function corresponding to the CE. The $\varphi_2(x)$ and $\varphi_3(x)$ functions describe the $\mathbf{F}_{4/4}$ and $\mathbf{F}_{8/12}$ boxes, correspondingly.

For example, for the $\mathbf{Q}_{2/1}$, $\mathbf{Q}_{4/4}$, and $\mathbf{Q}_{8/12}$ boxes we have, respectively,

$$\varphi_1(x) = 2^{-1}\,x + 2^{-1}\,x^2,\ \varphi_2(x) = 2^{-2}\,x + 3\cdot 2^{-3}\,x^2 + 2^{-2}\,x^3 + 2^{-3}\,x^4 \tag{3.25}$$

$$\begin{aligned}\varphi_3(x) &= 2^{-3}\,x + 7\cdot 2^{-5}\,x^2 + 7\cdot 2^{-5}\,x^3 + 25\cdot 2^{-7}\,x^4 \\ &+ 2^{-3}\,x^5 + 5\cdot 2^{-6}\,x^6 + 2^{-5}\,x^7 + 2^{-7}\,x^8\end{aligned} \tag{3.26}$$

Calculation results for the $\mathbf{Q}_{16/32}$, $\mathbf{Q}_{32/80}$, $\mathbf{Q}_{64/192}$, $\mathbf{R}_{32/80}$, and $\mathbf{R}_{64/192}$ boxes are presented in Tables 3.11 to 3.15.

The technique of the generating functions can be also applied to the $\mathbf{\Phi}_{n/m}$ boxes of the first order, although different variants of the $\mathbf{F}_{2/2}$ CEs are used as standard building blocks. For the A, B, C, D, and E types of the CEs $\mathbf{F}_{2/2}$, we have the following generating functions:

$$\text{A: } \varphi_1^A(x) = \frac{3}{4}x + \frac{1}{4}x^2 \tag{3.27}$$

$$\text{B: } \varphi_1^B(x) = \frac{5}{8}x + \frac{3}{8}x^2 \tag{3.28}$$

$$\text{C: } \varphi_1^C(x) = \frac{7}{8}x + \frac{1}{8}x^2 \tag{3.29}$$

$$\text{D: } \varphi_1^D(z) = \frac{1}{2}x + \frac{1}{2}x^2 \tag{3.30}$$

$$\text{E: } \varphi_1^E(x) = x \tag{3.31}$$

Table 3.11 Theoretic distribution of the probability $p(t) = \Pr(\Delta_t^Y/\Delta_1^X, \Delta_0^V)$ for the $\mathbf{Q}_{16/32}$ box

t	$p(t)$	t	$p(t)$	t	$p(t)$	t	$p(t)$
1	0.063	5	0.13	9	0.049	13	3.4×10^{-3}
2	0.12	6	0.12	10	0.032	14	1.1×10^{-3}
3	0.14	7	0.095	11	0.018	15	2.4×10^{-4}
4	0.15	8	0.071	12	8.5×10^{-3}	16	3.1×10^{-5}

Table 3.12 Theoretic distribution of the probability $p(t) = \Pr(\Delta_t^Y/\Delta_1^X, \Delta_0^V)$ for the $Q_{32/80}$ box

t	$p(t)$	t	$p(t)$	t	$p(t)$	t	$p(t)$
1	0.031	9	0.077	17	0.011	25	5.8×10^{-5}
2	0.061	10	0.067	18	6.8×10^{-3}	26	2.0×10^{-5}
3	0.076	11	0.057	19	4.2×10^{-3}	27	6.4×10^{-6}
4	0.090	12	0.041	20	2.5×10^{-5}	28	1.7×10^{-6}
5	0.093	13	0.037	21	1.4×10^{-3}	29	3.7×10^{-6}
6	0.096	14	0.029	22	7.0×10^{-4}	30	6.3×10^{-8}
7	0.091	15	0.021	23	3.3×10^{-4}	31	7.5×10^{-9}
8	0.085	16	0.015	24	1.5×10^{-4}	32	4.7×10^{-10}

Table 3.13 Theoretic distribution of the probability $p(t) = \Pr(\Delta_t^Y/\Delta_1^X, \Delta_0^V)$ for the $Q_{64/192}$ box

t	$p(t)$	t	$p(t)$	t	$p(t)$	t	$p(t)$
1	0.016	17	0.035	33	6.6×10^{-4}	49	2.3×10^{-8}
2	0.031	18	0.031	34	4.4×10^{-4}	50	8.6×10^{-9}
3	0.040	19	0.027	35	2.8×10^{-4}	51	3.0×10^{-9}
4	0.049	20	0.023	36	1.8×10^{-4}	52	10^{-9}
5	0.054	21	0.019	37	1.1×10^{-4}	53	3.1×10^{-10}
6	0.059	22	0.016	38	6.6×10^{-5}	54	8.8×10^{-11}
7	0.061	23	0.013	39	3.8×10^{-5}	55	2.3×10^{-11}
8	0.062	24	0.011	40	2.1×10^{-5}	56	5.6×10^{-12}
9	0.062	25	0.084	41	1.2×10^{-5}	57	1.2×10^{-12}
10	0.061	26	0.065	42	6.2×10^{-6}	58	2.3×10^{-13}
11	0.059	27	0.05	43	3.1×10^{-6}	59	3.9×10^{-14}
12	0.056	28	0.038	44	1.5×10^{-6}	60	5.6×10^{-15}
13	0.052	29	0.028	45	7.2×10^{-7}	61	6.5×10^{-16}
14	0.048	30	0.020	46	3.3×10^{-7}	62	5.7×10^{-17}
15	0.044	31	1.4×10^{-3}	47	1.4×10^{-7}	63	3.5×10^{-18}
16	0.040	32	9.8×10^{-4}	48	5.9×10^{-8}	64	1.1×10^{-18}

Table 3.14 Theoretic distribution of the probability $p(t) = \Pr(\Delta_t^Y/\Delta_1^X, \Delta_0^V)$ for the $R_{32/80}$ box

t	$p(t)$	t	$p(t)$	t	$p(t)$	t	$p(t)$
1	0.24	9	8.3×10^{-3}	17	5.2×10^{-6}	25	3.2×10^{-11}
2	0.24	10	4.0×10^{-3}	18	1.6×10^{-6}	26	4.5×10^{-12}
3	0.19	11	1.9×10^{-3}	19	4.4×10^{-7}	27	5.3×10^{-13}
4	0.13	12	8.1×10^{-4}	20	1.1×10^{-7}	28	5.3×10^{-14}
5	0.086	13	3.3×10^{-4}	21	2.1×10^{-8}	29	4.2×10^{-15}
6	0.052	14	1.3×10^{-4}	22	5.8×10^{-9}	30	2.6×10^{-16}
7	0.03	15	4.8×10^{-5}	23	1.2×10^{-9}	31	1.0×10^{-17}
8	0.016	16	1.6×10^{-5}	24	2.0×10^{-10}	32	2.2×10^{-19}

Table 3.15 Theoretic distribution of the probability $p(t) = \Pr(\Delta_t^Y/\Delta_1^X, \Delta_0^V)$ for the $R_{64/192}$ box

t	$p(t)$	t	$p(t)$	t	$p(t)$	t	$p(t)$
1	0.18	17	2.6×10^{-4}	33	1.5×10^{-10}	49	8.4×10^{-21}
2	0.2	18	1.3×10^{-4}	34	4.6×10^{-11}	50	1.2×10^{-21}
3	0.17	19	6.4×10^{-5}	35	1.4×10^{-11}	51	1.7×10^{-22}
4	0.14	20	3.1×10^{-5}	36	4.0×10^{-12}	52	2.2×10^{-23}
5	0.1	21	1.4×10^{-5}	37	1.1×10^{-12}	53	2.6×10^{-24}
6	0.074	22	6.5×10^{-6}	38	3.0×10^{-13}	54	2.9×10^{-25}
7	0.052	23	2.8×10^{-6}	39	7.8×10^{-14}	55	2.9×10^{-26}
8	0.034	24	1.2×10^{-6}	40	2.0×10^{-14}	56	2.6×10^{-27}
9	0.022	25	5.0×10^{-7}	41	4.6×10^{-15}	57	2.1×10^{-28}
10	0.014	26	2.0×10^{-7}	42	1.1×10^{-15}	58	1.4×10^{-29}
11	8.6×10^{-3}	27	8.0×10^{-8}	43	2.3×10^{-16}	59	8.7×10^{-31}
12	5.1×10^{-3}	28	3.0×10^{-8}	44	4.8×10^{-17}	60	4.4×10^{-32}
13	3.0×10^{-3}	29	1.1×10^{-8}	45	9.5×10^{-18}	61	1.8×10^{-33}
14	1.7×10^{-3}	30	4.0×10^{-9}	46	1.8×10^{-18}	62	5.5×10^{-35}
15	9.2×10^{-4}	31	1.4×10^{-9}	47	3.2×10^{-19}	63	1.1×10^{-36}
16	4.9×10^{-4}	32	4.6×10^{-10}	48	5.3×10^{-20}	64	1.2×10^{-38}

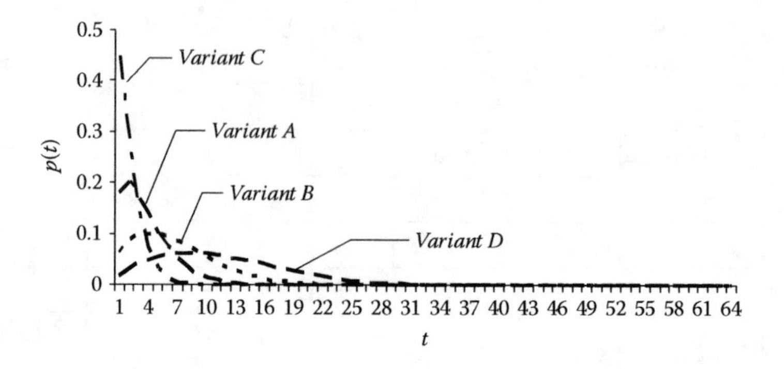

Figure 3.11 Distribution $p(t) = \Pr(\Delta_t^Y/\Delta_1^X, \Delta_0^V)$ for the $\Phi_{64/384}$ boxes (calculated using the generated functions technique).

Using these initial generating functions, we have performed calculations for the first order boxes $\mathbf{\Phi}_{64/384}$ with symmetric and nonsymmetric topologies. The obtained results are shown in Figure 3.11. The performed statistic experiments confirm the theoretic calculations.

3.5 Data-Driven Ciphers Based on CSPNs

3.5.1 Block Cipher Eagle-128

The 128-bit block cipher Eagle-128 has been proposed for application in the cases where restricted hardware resources are available for embedded cryptographic modules using the FPGAs and application-specific integrated circuits (ASIC) technologies [46]. Hardware efficacy of Eagle-128 is provided by (1) combining the CSPNs with SPNs in the round transformation, (2) using the advanced $\mathbf{F}_{2/2}$ CEs as elementary building blocks of the CSPNs, and (3) simultaneously transforming the whole data block in one round. The $\mathbf{\Phi}_{64/384}$ and $\mathbf{\Phi}_{32/32}$ operations built up using the $\mathbf{F}_{2/2}$ = (e,i,g,f) and $\mathbf{F}'_{2/2}$ = (e,b,b,c) elements, respectively, as standard building blocks represent the CSPN part of the cipher. The SPN part is implemented as two mutually inverse networks built up using two layers of the 4 × 4 S boxes. Each of the layers contains eight S boxes. The $\mathbf{\Phi}_{64/384}$ box is implemented as uniform CSPN having the first order topology, in which the active layer contains 32 elements (e,i,g,f). This concrete variant of the CEs was selected as elementary primitive because the algebraic degree of its BFs f_1, f_2, and f_3 is equal to 3 (this is the maximum value for the $\mathbf{F}_{2/2}$ elements). The outputs of the (e,i,g,f) elements are described with the following BFs:

$$f_1 = y_1 = vzx_2 \oplus vx_2 \oplus vx_1 \oplus zx_1 \oplus z \oplus x_2;\ \mathrm{NL}(y_1) = 4 \tag{3.32}$$

$$f_2 = y_2 = vzx_1 \oplus vz \oplus vx_2 \oplus zx_1 \oplus zx_2 \oplus x_1;\ \mathrm{NL}(y_2) = 4 \tag{3.33}$$

$$f_3 = y_1 \oplus y_2 = vzx_1 \oplus vzx_2 \oplus vz \oplus vx_1 \oplus zx_2 \oplus z \oplus x_1 \oplus x_2;\ \mathrm{NL}(y_1 \oplus y_2) = 4 \tag{3.34}$$

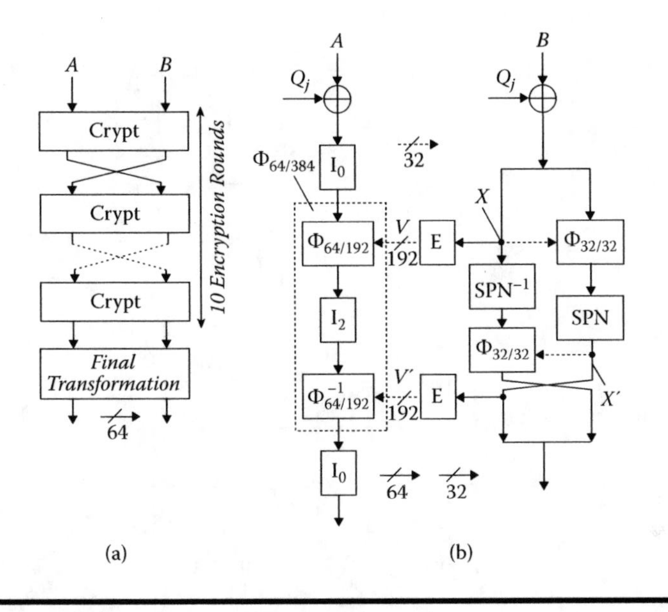

Figure 3.12 General structure of Eagle-128 (*a*) and specification of the round transformation Crypt (*b*).

In the $\mathbf{\Phi}_{32/32}$ boxes, the (e,b,b,c) elements are used. This variant of the $\mathbf{F}_{2/2}$ elements has been selected to provide strengthening of the diffusion property of the $\mathbf{\Phi}_{32/32}$ box. This type of CEs is described by the following BFs:

$$y_1 = vzx_1 \oplus vzx_2 \oplus vx_1 \oplus vx_2 \oplus zx_1 \oplus zx_2 \oplus z \oplus v \oplus x_2;\ \mathrm{NL}(y_1) = 2 \quad (3.35)$$

$$y_2 = vzx_1 \oplus vzx_2 \oplus vz \oplus vx_1 \oplus vx_2 \oplus zx_1 \oplus zx_2 \oplus x_1;\ \mathrm{NL}(y_2) = 2 \quad (3.36)$$

$$y_3 = vz \oplus v \oplus z \oplus x_1 \oplus x_2;\ \mathrm{NL}(y_3) = 4 \quad (3.37)$$

Changing a single bit at the controlling input of the $\mathbf{\Phi}_{32/32}$ box causes deterministically some change at its output, namely, one or two output bits are inverted (each of these two events has probability 0.5). Figure 3.12*a* shows the general iterative structure of Eagle-128, where the **Crypt** box denotes the round transformation procedure. The full encryption procedure in Eagle-128 is performed as follows:

1. For $j = 1$ to 9, do: $\{(L, R) \leftarrow \mathbf{Crypt}^{(e)}(L, R, Q_j); (L, R) \leftarrow (R, L)\}$.
2. Perform transformation: $\{(L, R) \leftarrow \mathbf{Crypt}^{(e)}(L, R, Q_{10})\}$.
3. Perform final transformation: $\{(L, R) \leftarrow (L \oplus Q_{11}, R \oplus Q_{11})\}$.

The structure of the encryption round is presented in Figure 3.12*b*, where the $\mathbf{\Phi}_{64/384}$ box is represented as superposition $\mathbf{\Phi}_{64/192} \circ \mathbf{I}_2 \circ \mathbf{\Phi}^{-1}_{64/192}$. The permutational involution $\mathbf{I}_2$ and detailed topology of the $\mathbf{\Phi}_{64/384}$ box are specified in Section 3.3. Each of two mutually inverse switchable permutation networks (SPNs) used in the right branch (see Figure 3.13) represents two active layers of the 4×4 substitutions.

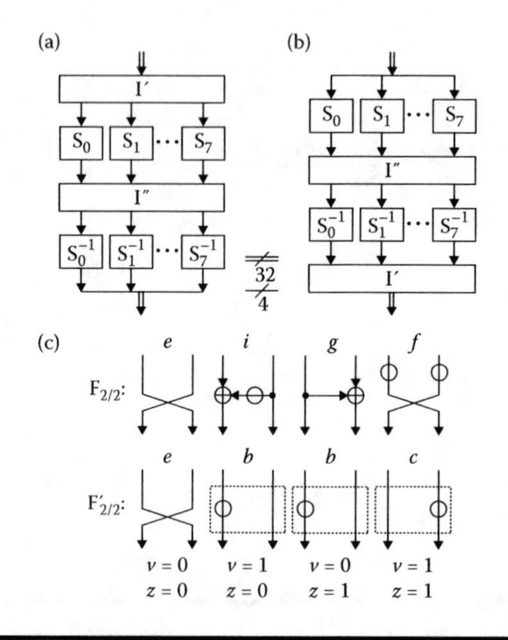

Figure 3.13 Specification of mutually inverse operations SPN (*a*) and SPN^{-1} (*b*), and modifications of the elementary transformations implemented by the $F_{2/2}$ and $F'_{2/2}$ controlled elements (*c*).

The upper layer includes the $S_0,\ldots,S_7$ boxes specified in Table 3.16. The lower layer includes the $S_0^{-1}, \ldots, S_7^{-1}$ boxes specified in Table 3.17. Eight 4×4 S boxes of the DES cipher (one from each of its eight 6×4 S boxes) have been selected as the $S_0, \ldots, S_7$ boxes of Eagle-128 to inspire a high level of public confidence that no trapdoors are inserted. The permutations **I′** and **I″** are described as follows:

I′: (1)(2,18)(3)(4,20)(5)(6,22)(7)(8,24)(9)(10,26)(11)(12,28)
(13)(14,30)(15)(16,32)(17)(19)(21)(23)(25)(27)(29)(31)

and

I″: (1)(2,5)(3,9)(4,13)(6)(7,10)(8,14)(11)(12,15)(16)(17)
(18,21)(19,25)(20,29)(22)(23,26)(24,30)(27)(28,31)(32)

The permutation $\mathbf{I}_0$ is used to prevent two output bits of the same $\mathbf{F}_{2/2}$ element of the $\mathbf{\Phi}_{64/384}$ box to be fed to the input of the same 4×4 S box or to the input of the same CE of the $\mathbf{\Phi}_{32/32}$ box. The permutation $\mathbf{I}_0$ is an involution described as follows:

$$(1)(2, 34)\ldots(2i-1)(2j, 2j + 32)\ldots(63)(32, 64),$$

where $i = 1, 2, \ldots, 32$ and $j = 1, 2, \ldots, 16$.

Table 3.16 Specification of the 4×4 substitution boxes S_0, ..., S_7 (input and output four-bit binary vectors are presented in the form of binary numbers)

Input value		0	1	2	3	4	5	6	7	8	9	10	11	12	13	14	15
Output value	S_0	14	4	13	1	2	15	11	8	3	10	6	12	5	9	0	7
	S_1	3	13	4	7	15	2	8	14	12	0	1	10	6	9	11	5
	S_2	10	0	9	14	6	3	15	5	1	13	12	7	11	4	2	8
	S_3	1	4	11	13	12	3	7	14	10	15	6	8	0	5	9	2
	S_4	10	6	9	0	12	11	7	13	15	1	3	14	5	2	8	4
	S_5	11	8	12	7	1	14	2	13	6	15	0	9	10	4	5	3
	S_6	10	15	4	2	7	12	9	5	6	1	13	14	0	11	3	8
	S_7	1	15	13	8	10	3	7	4	12	5	6	11	0	14	9	2

Table 3.17 Specification of the 4 × 4 substitution boxes S_0^{-1}, ..., S_7^{-1}

Input value		0	1	2	3	4	5	6	7	8	9	10	11	12	13	14	15
Output value	S_0^{-1}	14	3	4	8	1	12	10	15	7	13	9	6	11	2	0	5
	S_1^{-1}	9	10	5	0	2	15	12	3	6	13	11	14	8	1	7	4
	S_2^{-1}	1	8	14	5	13	7	4	11	15	2	0	12	10	9	3	6
	S_3^{-1}	12	0	15	5	1	13	10	6	11	14	8	2	4	3	7	9
	S_4^{-1}	3	9	13	10	15	12	1	6	14	2	0	5	4	7	11	8
	S_5^{-1}	10	4	6	15	13	14	8	3	1	11	12	0	2	7	5	9
	S_6^{-1}	12	9	3	14	2	7	8	4	15	6	0	13	5	10	11	1
	S_7^{-1}	12	0	15	5	7	9	10	6	3	14	4	11	8	2	13	1

The Eagle-128 algorithm uses the 256-bit secret key $K = (K_1, K_2, K_3, K_4)$, where subkeys $K_i \in \{0, 1\}^{64}$ are used directly in procedure **Crypt** as round keys Q_j (encryption) or Q'_j (decryption) specified in Table 3.18. Because no preprocessing of the secret key is performed, Eagle-128 is fast in the cases when the keys are changed frequently. In each encryption round, only one of the 64-bit subkeys is used. It is combined with both the left and the right data subblocks. Procedure **Crypt** is not involution, its part after combining the round key with data subblocks is involution,

Table 3.18 Key scheduling

Round number j =	1	2	3	4	5	6	7	8	9	10	11
Encryption Q_j =	K^1	K^2	K^3	K^4	K^2	K^1	K^3	K^4	K^3	K^2	K^1
Decryption Q'_j =	K^1	K^2	K^3	K^4	K^3	K^1	K^2	K^4	K^3	K^2	K^1

Note: $j = 11$ corresponds to final transformation.

though. A very simple final transformation (FT) is used to symmetrize the full ten-round encryption procedure. The FT is implemented as XORing a subkey with both data subblocks. Due to FT in Eagle-128, the same algorithm performs both the encryption and the decryption, although different key scheduling is used.

The 192-bit controlling vectors V and V' corresponding to the $\mathbf{\Phi}_{64/192}$ and $\mathbf{\Phi}^{-1}_{64/192}$ boxes are formed with the extension box $\mathbf{E}$, described as follows:

$$\mathbf{E}(X) = V = (V_1, Z_1, V_2, Z_2, V_3, Z_3);\ V_i = X^{<<<10(i-1)};\ Z_i = X^{<<<10i-5};\ i = 1, 2, 3 \quad (3.38)$$

The 32-bit controlling vector (V_1, Z_1) of the $\mathbf{\Phi}_{32/32}$ operation is described as follows:

$$V_1 = (x_1, \ldots, x_{16}) \text{ and } Z_1 = (x_{17}, \ldots, x_{32}).$$

Statistic properties of Eagle-128 have been investigated using standard tests proposed by Preneel and others [54] for testing block ciphers. Three rounds of Eagle-128 are sufficient to satisfy the test criteria. Thus, Eagle-128 possesses good statistical properties similar to those of AES finalists. Security estimation shows that a differential attack against Eagle-128 is more efficient than a linear attack. Four (eight) rounds are sufficient to thwart linear (differential) attack. The best DCs are presented in Table 3.19, where (Δ_h^A, Δ_z^B) and $(\Delta'^A_{h'}, \Delta'^B_{z'})$ denote input and output differences, respectively. The probability that the difference (Δ_h^A, Δ_z^B) transforms into the difference $(\Delta'^A_{h'}, \Delta'^B_{z'})$ while passing through r rounds is denoted as $P(r)$. The value $P'(r)$ denotes the contribution of the considered mechanisms of the DC formation to the $P(r)$ value. The last column corresponds to the experiment evaluation of $P(r)$, i.e., to the case when all of the DC formation mechanisms are taken into account.

The probability to have at output of the random cipher the difference $(\Delta'^A_2, \Delta'^B_0)$ is equal to $P_{\text{rand}} > 2^{-117} > 2^{-124} \geq P'(8)$. Thus, the cipher Eagle-128 with eight encryption rounds appears to be indistinguishable from a random cipher with the most efficient DCs.

3.5.2 DDO-64: A DDO-Based Cipher with 64-Bit Data Block

The main primitives in the DDO-64 algorithm are the uniform boxes $\mathbf{\Phi}_{16/64}$ and $\mathbf{\Phi}^{-1}_{16/64}$ comprising four active layers in which the CE $\mathbf{F}_{2/2}$ = (h,f,e,j) is used as typical

Table 3.19 Efficient DCs of Eagle-128

Input difference	Output difference	R	$P'(r)$ [46]	$P(r)$ (experiment)
(Δ_2^A, Δ_0^B)	$(\Delta_4'^A, \Delta_0'^B)$	2	$\approx 2^{-38.5}$	$< 2^{-39}$
(Δ_4^A, Δ_0^B)	$(\Delta_2'^A, \Delta_0'^B)$	2	$\approx 2^{-39}$	$< 2^{-31}$
(Δ_2^A, Δ_0^B)	$(\Delta_2'^A, \Delta_0'^B)$	2	$\approx 2^{-36.5}$	$< 2^{-31}$
(Δ_2^A, Δ_0^B)	$(\Delta_2'^A, \Delta_0'^B)$	4	$\approx 2^{-73}$	—
(Δ_2^A, Δ_0^B)	$(\Delta_2'^A, \Delta_0'^B)$	6	$\approx 2^{-109.5}$	—
(Δ_2^A, Δ_0^B)	$(\Delta_2'^A, \Delta_0'^B)$	8	$\approx 2^{-145}$	—
(Δ_2^A, Δ_0^B)	$(\Delta_0'^A, \Delta_2'^B)$	10	$\approx 2^{-180}$	—

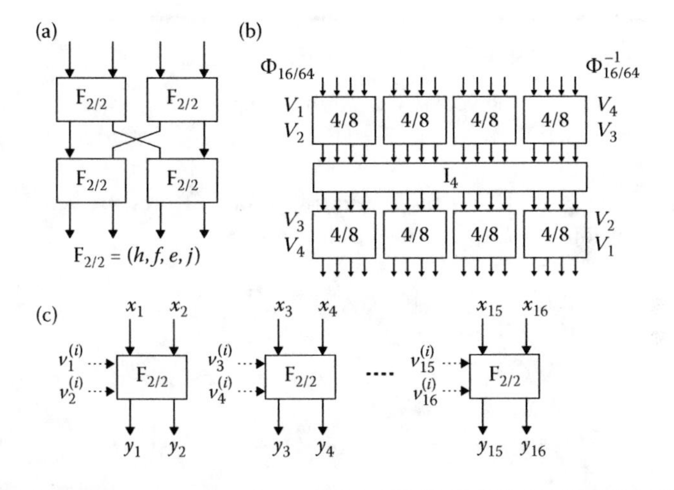

Figure 3.14 The $\Phi_{4/8}$ (*a*), $\Phi_{16/64}$, $\Phi^{-1}_{16/64}$ (*b*) boxes and the controlling bits distribution in the *i*-th layer (*c*).

building element. Both of these mutually inverse operations are constructed using the $\mathbf{\Phi}_{4/8}$ boxes (see Figures 3.14*a*, *b*, and *c*). The $\mathbf{\Phi}_{16/64}$ and $\mathbf{\Phi}^{-1}_{16/64}$ boxes have the same symmetric topology represented structurally as two cascades of the $\mathbf{\Phi}_{4/8}$ boxes that are connected with the permutational involution $\mathbf{I}_4$. The difference between them consists in different distribution of the controlling bits. The permutation $\mathbf{I}_4$ connects each of four upper boxes $\mathbf{\Phi}_{4/8}$ with each of lower ones providing mirror symmetry of the $\mathbf{\Phi}_{16/64}$ box structure. The $\mathbf{I}_4$ box is described as follows:

$$(1)(2,5)(3,9)(4,13)(6)(7,10)(8,14)(11)(12,15)(16).$$

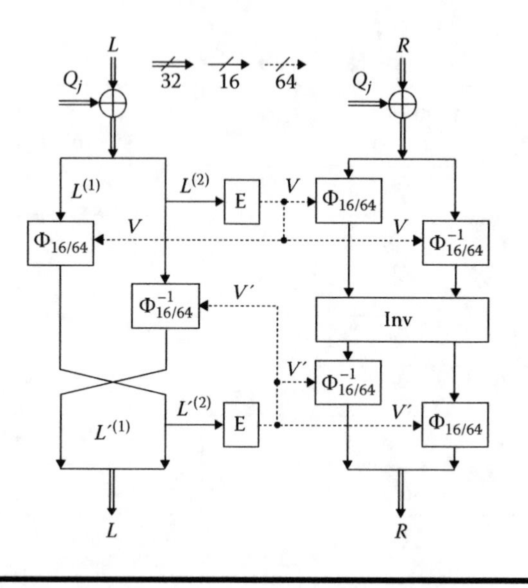

Figure 3.15 Structure of the round transformation Crypt.

We assume that controlling vector V contains s components, where s is the number of the layers in the DDO box, i.e., $V = (V^{(1)}, V^{(2)}, \ldots, V^{(s)})$. Distribution of bits of $V^{(i)} = (v_1^{(i)}, \ldots, v_{16}^{(i)})$, $i = 1, \ldots, s$, is presented in Figure 3.14*c*.

Specification of the DDO-64 encryption round is presented in Figure 3.15, where the extension box **E** is described as follows:

$$\mathbf{E}(L^{(2)}) = V = (V^{(1)}, \ldots, V^{(4)}),\ L^{(2)} = (l_1, \ldots, l_{16}) \tag{3.39}$$

$$\forall\ i \in \{1,2,3,4\}:\ v_1^{(i)} = v_5^{(i)} = v_9^{(i)} = v_{13}^{(i)} = l_{4i-3};\ v_2^{(i)} = v_6^{(i)} = v_{10}^{(i)} = v_{14}^{(i)} = l_{4i-2}$$

$$v_3^{(i)} = v_7^{(i)} = v_{11}^{(i)} = v_{15}^{(i)} = l_{4i-1};\ v_4^{(i)} = v_8^{(i)} = v_{12}^{(i)} = v_{16}^{(i)} = l_{4i} \tag{3.40}$$

The fixed permutation **Inv** represents the following involution:

(1,17)(2,21)(3,25)(4,29)(5,18)(6,22)(7,26)(8,30)(9,19)
(10,23)(11,27(12,31)(13,20)(14,24)(15,28)(16,32).

The encryption algorithm includes the following generalized steps:

1. For $j = 1$ to 7, do: $\{(L, R) \leftarrow$ **Crypt**$(L, R, Q_j);\ (L, R) \leftarrow (R, L)\}$.
2. Transform the (L, R) data block: $(L, R) \leftarrow$ **Crypt**(L, R, Q_8).
3. Perform final transformation: $\{(L, R) \leftarrow (L \oplus Q_9, R \oplus Q_9)\}$.

The DDO-64 performs with 128-bit secret key $K = (K_1, K_2, K_3, K_4)$, the subkeys of which $K_i \in \{0, 1\}^{32}$ are used directly in the round transformation as round keys

Table 3.20 The key scheduling in DDO-64

j =	1	2	3	4	5	6	7	8	9
Q_j = (*encryption*)	K_1	K_2	K_3	K_4	K_1	K_2	K_4	K_3	K_1
Q'_j = (*decryption*)	K_1	K_3	K_4	K_2	K_1	K_4	K_3	K_2	K_1

Q_j (while encrypting) or Q'_j (while decrypting). The key scheduling is presented in Table 3.20, where the subkeys used to perform the final transformation correspond to $j = 9$.

The $\mathbf{\Phi}_{16/64}$ and $\mathbf{\Phi}^{-1}_{16/64}$ boxes provide sufficiently better avalanche than the CP boxes, because changing one bit at their input causes on the average changing many bits. Besides, each bit of the left data subblock is used to control twelve different CEs in three DDO boxes. We can get some strengthening of the avalanche by adding two permutational involutions $\mathbf{I}'_0$ in the right branch of the round transformation—one just after the XOR operation and one in the end of the right branch—as in the left branch of the Eagle-128 cipher described in the previous paragraph. One of the possible variants of the $\mathbf{I}'_0$ involution is $\mathbf{I}'_0 = \mathbf{Inv}$. However, in the initially proposed version of DDO-64 the $\mathbf{I}'_0$ involution is not used [40].

Security estimation of the DDO-64 shows that six rounds are sufficient to thwart differential attacks and four rounds are sufficient to thwart linear cryptanalysis. The best DCs have probability $P(2) < 2^{-29}$ and correspond to the two-round iterative differences with one active bit. For the random 64-bit cipher, such DCs have probability $2^{-59} >> P^3(2)$, where $P^3(2)$ is the contribution of the mentioned iterative DCs to the probability of the six-round DDO-64.

The DDO-64 is very efficient for the FPGA and ASIC hardware implementations. It can operate at high frequencies because the time delay τ of the round transformation is sufficiently small (when the FPGA implementation $\tau = 9t_{\oplus}$, where $t_{\oplus}$ is the time delay of the XOR operation).

3.5.3 Updating the Known DDP-Based Ciphers

In the known DDP-based block ciphers, all of the $\mathbf{P}_{2/1}$ switching elements can be replaced with the single type elements that correspond to the class $\mathbf{F}_{2/1}$ or $\mathbf{F}_{2/2}$ (this relates to the use of the respective DDO boxes instead of the DDP boxes). This will improve the linear and differential properties of the DDOs, and a smaller number of the encryption rounds will be required for secure encryption. Such approach has been applied to the following DDP-based ciphers [40]: Cobra-H64, Cobra-H128, and Spectr-H64. Investigated variants of the modified ciphers are presented in Table 3.21. The security estimation results against differential attack are presented in Table 3.22, where $P(N)$ denotes the probability with which the iterative differences pass through N rounds. We see that replacement of the DDP operations by

Table 3.21 Updating some DDP-based ciphers

Cipher	Data block size, bits	$t_{\oplus}$[a], arb.un.	DDO boxes		Secure number of rounds	
			Initial	New[b]	Initial	New
Cobra-H64 [65]	64	**15**	$\mathbf{P}_{32/96}$ $\mathbf{P}^{-1}_{32/96}$	$\mathbf{R}_{32/96}$ $\mathbf{R}^{-1}_{32/96}$	**10**	**8**
Cobra-H64 [65]	64	**15**	$\mathbf{P}_{32/96}$ $\mathbf{P}^{-1}_{32/96}$	$\mathbf{\Phi}_{32/192}$ $\mathbf{\Phi}^{-1}_{32/192}$	**10**	**6**
Cobra-H128 [65]	128	**14**	$\mathbf{P}_{64/192}$ $\mathbf{P}^{-1}_{64/192}$	$\mathbf{Q}_{64/192}$ $\mathbf{Q}^{-1}_{64/192}$	**12**	**9**
Cobra-H128 [65]	128	**14**	$\mathbf{P}_{64/192}$ $\mathbf{P}^{-1}_{64/192}$	$\mathbf{\Phi}_{64/384}$ $\mathbf{\Phi}^{-1}_{64/384}$	**12**	**6**
Spectr-H64 [21,8]	64	**12**	$\mathbf{P}_{32/80}$ $\mathbf{P}^{-1}_{32/80}$	$\mathbf{R}_{32/80}$ $\mathbf{R}^{-1}_{32/80}$	**12**	**8**

[a] Time delay of the round transformation.

[b] The **R**-, **Q**-, and **Φ**-boxes are constructed using the (i,e), (i,j), and (h,f,e,j) CEs, respectively.

Table 3.22 Security comparison against DCA

Cipher	N (number of rounds)	DDO boxes	*P*(2)	*P*(*N*)
Cobra-H64	**10**	**P**-type [42]	$\mathbf{1.16 \cdot 2^{-19}}$	$\mathbf{\approx 2^{-94}}$
	8	**R**-type	$\mathbf{< 2^{-24}}$	$\mathbf{< 2^{-96}}$
	6	**Φ**-type	$\mathbf{< 2^{-36}}$	$\mathbf{< 2^{-108}}$
Cobra-H128	**12**	**P**-type [65]	$\mathbf{1.13 \cdot 2^{-29}}$	$\mathbf{\approx 2^{-173}}$
	9	**Q**-type	$\mathbf{< 2^{-46}}$	$\mathbf{< 2^{-184}}$
	6	**Φ**-type	$\mathbf{< 2^{-57}}$	$\mathbf{< 2^{-171}}$
Spectr-H64	**12**	**P**-type [8]	$\mathbf{1.15 \cdot 2^{-13}}$	$\mathbf{< 2^{-77}}$
	8	**R**-type	$\mathbf{< 2^{-20}}$	$\mathbf{< 2^{-80}}$
DDO-64	**6**	**Φ**-type	$\mathbf{< 2^{-29}}$	$\mathbf{< 2^{-87}}$

the DDO ones significantly reduces the value of $P(2)$ and the minimum number of the encryption rounds required to get the value $P(N) < 2^{-59}$ for 64-bit ciphers and the value $P(N) < 2^{-122}$ for 128-bit ones (these figures correspond to the probabilities of the considered DCs in the case of random transformation of the 64-bit and 128-bit data blocks, respectively). Security comparison of the initial and different variants of the modified ciphers is shown in Table 3.22.

3.6 Conclusions

Results of this chapter can be summarized as follows:

- Two new classes, $\mathbf{F}_{2/1}$ and $\mathbf{F}_{2/2}$, of the minimum-size CEs have been introduced to design different types of DDOs oriented to application in the design of the fast ciphers suitable to cheap FPGA and ASIC implementation.
- Classification of the CEs in each of the classes $\mathbf{F}_{2/1}$ and $\mathbf{F}_{2/2}$ has been performed.
- The notion of the symmetric topology of the CSPN has been introduced. Several variants of the symmetric topologies of different orders have been proposed.
- Several types of the DDO boxes have been applied to update some known DDP-based block encryption systems. Using the DDO boxes instead of the analogous DDP boxes provides secure reduction of the number of encryption rounds, i.e., contributes to faster encryption in the cipher block chaining (CBC) mode.
- New block ciphers DDO-64 and Eagle-128 suitable to efficient FPGA and ASIC implementation have been described.

Chapter 4

Switchable Data-Dependent Operations

First, it is shown that the controlled permutation (CP) boxes constructed using recursive mechanisms of the first, second, and third types implement the sets that bit permutations each of which can be represented as a set of pairs of mutually inverse fixed bit permutations. Then, this chapter introduces the notion of the switchable controlled operations (SCOs), and different methods of their synthesis are presented. Topologies of the SCOs of different orders are described. Architecture of the bit permutation instruction oriented to both cryptographic and noncryptographic applications, which is of interest for imbedding in general purpose processors, is proposed. The SCOs are discussed as cryptographic primitive. Several data-driven block ciphers based on SCOs are described.

4.1 Representation of the CP Boxes as a Set of Pairs of Mutually Inverse Modifications

4.1.1 Topologies of the First Order

Describing different topologies of the CP boxes, we have indicated that CPs with symmetric structure are of special interest. Indeed, symmetric, mutually inverse CP boxes have the same topology. They differ only in distribution of the controlling bits. A corollary from this fact is the following statement.

Statement 4.1. *Suppose* $\mathbf{P}_{n/m}$ *and* $\mathbf{P}^{-1}_{n/m}$ *are mutually inverse CP boxes having mirror symmetry topology. Then, each fixed bit permutation modification* Π_V *implemented with the* $\mathbf{P}_{n/m}$ *box can be implemented also with the* $\mathbf{P}^{-1}_{n/m}$ *box as fixed bit permutation modification* $\Pi'_{V'}$.

The proof of this statement is sufficiently evident. Suppose $V = (V_1, V_2, \ldots, V_s)$. Due to the mirror symmetry of the $\mathbf{P}_{n/m}$ topology, this CP box implements two mutually inverse fixed bit permutation modifications, Π_V and $\Pi_{V'}$, where $V' = (V_s, V_{s-1}, \ldots, V_1)$, i.e., $\Pi_V = (\Pi_{V'})^{-1}$. The bit permutation modification $\Pi'_{V'}$ implemented with the $\mathbf{P}^{-1}_{n/m}$ box is the inverse of $\Pi_{V'}$, i.e., $\Pi'_{V'} = (\Pi_{V'})^{-1} = \Pi_V$.

Thus, the $\mathbf{P}_{n/m}$ and $\mathbf{P}^{-1}_{n/m}$ boxes generate the same set of bit permutation modifications and define mutual inverse modifications for the given controlling vector V. This means that for all $V \in \{0, 1\}^m$, there exists $V' \in \{0, 1\}^m$ such that Π_V and $\Pi_{V'}$ modifications implemented with the $\mathbf{P}_{n/m}$ box (or with the $\mathbf{P}^{-1}_{n/m}$ box) are mutual inverses. In other words, the symmetric CP boxes can be represented as a set of pairs of mutually inverse modifications. It is unexpected, but some of the CP boxes having nonsymmetric topology possess the same property. In this paragraph, we will show that an arbitrary box having recursive topology of the first order represents an example of such CP boxes. Let us consider the first type of the recursive construction of the CP boxes and transpose two switching elements in the $\mathbf{P}_{8/12}$ box conserving their connections with the controlling bits and with the $\mathbf{P}_{2/1}$ elements in the upper and lower active layers. This topological transformation is presented in Figures 4.1*a* and *b*. We see that the topology of the first type recursive construction has been transformed into the topology of the second type recursive construction that gives the topology of the $\mathbf{P}^{-1}_{8/12}$ box. However, the transformation has not changed the Π_V modification for the arbitrary controlling vector because we have transposed the switching elements retaining their internal connections, i.e., the new box is the initial one. Comparing these two representations of the $\mathbf{P}_{8/12}$ box, we can conclude that for arbitrary vector $V = (v_1, v_2, \ldots, v_{12})$, Π_V and $\Pi_{V'}$ modifications implemented with the $\mathbf{P}_{8/12}$ box are mutual inverses if $V' = (v'_1, v'_2, \ldots, v'_{12})$ is such that $v'_1 = v_9$, $v'_2 = v_{10}$, $v'_3 = v_{11}$, $v'_4 = v_{12}$, $v'_5 = v_5$, $v'_6 = v_7$, $v'_7 = v_6$, $v'_8 = v_8$, $v'_9 = v_1$, $v'_{10} = v_2$, $v'_{11} = v_3$, and $v'_{12} = v_4$. In other words, the set of the bit permutation modifications implemented with the $\mathbf{P}_{8/12}$ box contains Π_V modification if, and only if, it contains the modification $\Pi_{V'} = \Pi_V^{-1}$. For symmetric CP boxes analogous fact can be formulated as the following corollary from Statement 4.1:

Corollary 4.1. *Arbitrary CP boxes having mirror symmetry topology implement a set of fixed bit permutations that can be represented as a set of pairs of mutual inverses* Π_V *and* $\Pi_{V'} = \Pi_V^{-1}$.

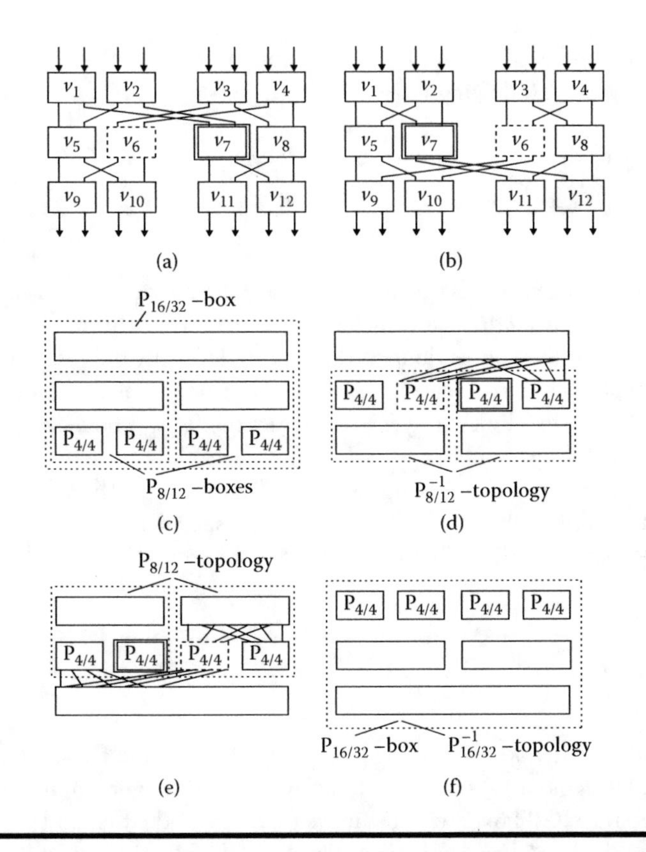

Figure 4.1 Topological transformations: (*a*, *b*) representation of the $P_{8/12}$ box in two mutual inverse topologies; (*c*) representation of the $P_{16/32}$ box in recursive topology of the first type; (*d*, *e*) representation of the $P_{16/32}$ box in combined topology; (*f*) the $P_{16/32}$ box represented in recursive topology of the second type.

A statement such as this can be formulated and proved for CP boxes with non-symmetric topology. For example, the following statement is valid for the first order CP boxes having recursive structure.

Statement 4.2. *Each of the first order boxes* $\mathbf{P}_{n/m}$ *and* $\mathbf{P}^{-1}_{n/m}$ *having recursive structure implements a set of fixed bit permutations that can be represented as a set of pairs of mutual inverses* Π_V *and* $\Pi_{V'} = \Pi_V^{-1}$.

We will consider only the proof of Statement 4.2. Let an unbroken band denote an active layer linking two CP boxes $\mathbf{P}_{n/m}$ of the first order in a larger box $\mathbf{P}_{2n/2m+n}$; let us consider the recursive construction mechanism of the first type (see Chapter 2, Section 2.1). It is assumed that $n = 2^z$, where $z > 2$ is a natural number. The fixed bit permutation π_1 defining connections of the top (first) linking active layer with the second active layer is called linking fixed permutation $\pi_{(z+1)}$. The $\pi_{(z+1)}$ permutation is described with the following table:

$$\begin{pmatrix} 1 & 2 & 3 & 4 & \dots & 2i-1 & 2i & \dots & 2n-1 & 2n \\ 1 & n+1 & 2 & n+2 & \dots & i & n+i & \dots & n & 2n \end{pmatrix} \tag{4.1}$$

where the upper row indicates the initial position of the permuted bits and the lower row indicates the output digits to which the bits are moved. This table gives the general description of the connections of the linking layers in the recursive topology of the first type. While considering recursive construction of the second type, the bottom layer serves as a linking layer to construct the $\mathbf{P}^{-1}_{2n/2m+n}$ box. The fixed bit permutation π_{s-1} defining connections of the bottom (sth) linking active layer with the (s – 1)th active layer is called linking fixed permutation $\pi^{-1}_{(z+1)}$. The $\pi^{-1}_{(z+1)}$ permutation is described with the following table giving general specification of the connections of the linking layers in the recursive topology of the second type:

$$\begin{pmatrix} 1 & 2 & 3 & 4 & \dots & i & \dots & n & n+1 & \dots & n+i & \dots & 2n-1 & 2n \\ 1 & 3 & 5 & 7 & \dots & 2i-1 & \dots & 2n-1 & 2 & \dots & 2i & \dots & 2n-2 & 2n \end{pmatrix} \tag{4.2}$$

Let $\mathbf{L}_{(z+1)}$ and $\mathbf{L}^{-1}_{(z+1)}$ denote the linking layers in the $\mathbf{P}_{2n/2m+n}$ and $\mathbf{P}^{-1}_{2n/2m+n}$ boxes, correspondingly. Using notations of the linking layers and linking permutations, we can represent arbitrary CP boxes with recursive topology of the first and second types. For example, the $\mathbf{P}_{16/32}$ box can be represented as the following superposition:

$$\begin{aligned} \mathrm{P}_{16/32} = {} & \mathrm{L}_{(4)} \circ \pi_{(4)} \circ \left(\mathrm{L}_{(3)} \,||\, \mathrm{L}_{(3)}\right) \circ \left(\pi_{(3)} \,||\, \pi_{(3)}\right) \circ \left(\mathrm{L}_{(2)} \,||\, \mathrm{L}_{(2)} \,||\, \mathrm{L}_{(2)} \,||\, \mathrm{L}_{(2)}\right) \circ \\ & \circ \left(\pi_{(2)} \,||\, \pi_{(2)} \,||\, \pi_{(2)} \,||\, \pi_{(2)}\right) \circ \\ & \circ \left(\mathrm{P}_{2/1} \,||\, \mathrm{P}_{2/1} \,||\, \mathrm{P}_{2/1} \,||\, \mathrm{P}_{2/1} \,||\, \mathrm{P}_{2/1} \,||\, \mathrm{P}_{2/1} \,||\, \mathrm{P}_{2/1} \,||\, \mathrm{P}_{2/1}\right) \end{aligned} \tag{4.3}$$

where $\mathbf{F}||\mathbf{F}'$ denotes concatenation of the transformations $Y = \mathbf{F}(X)$ and $Y' = \mathbf{F}'(X')$, defined as follows: $Y||Y' = \mathbf{F}||\mathbf{F}'(X||X')$.

Using schematic representation of the linking layers, the $\mathbf{P}_{16/32}$ box can be figured as is shown in Figure 4.1*c*. Transposing respective two $\mathbf{P}_{2/1}$ elements in each of the $\mathbf{P}_{8/12}$ boxes as illustrated in Figures 4.1*a* and *b*, we represent the lasts as the boxes with inverse topology (i.e., with recursive topology of the second type with which the recursive boxes $\mathbf{P}^{-1}_{n/m}$ are constructed). This leads to representation of $\mathbf{P}_{16/32}$ box with a topology combining the first and the second recursive construction mechanisms as is shown in Figure 4.1*d*. Actually, we have the same box as the initial one because the topological transformations are accompanied with respective changes in the distribution of the controlling bits. Then, we perform the next identical topologic transformation as transposition of two inner $\mathbf{P}_{4/4}$ boxes, which gives the representation of the $\mathbf{P}_{16/32}$ box in the combined topology shown in Figure 4.1*e*. Now we have two individual topologies of the first type linked with the linking layer of the second type. Each of these $\mathbf{P}_{8/12}$ topologies can be transformed to the $\mathbf{P}^{-1}_{8/12}$ topologies. Thus, we have the representation of the $\mathbf{P}_{16/32}$ box in the recursive topology of the second type ($\mathbf{P}^{-1}_{16/32}$ topology shown in Figure 4.1*f*). Actually, in the initial and the last representations of the $\mathbf{P}_{16/32}$ box, we have different distributions of active bits, but in both cases, for given controlling vector V, the same bit permutation modification Π_V is implemented. Because the considered representations correspond to mutually inverse topologies, for arbitrary given V, there exists V' such that $\Pi_{V'} = \Pi_V^{-1}$.

Suppose, for arbitrary $n = 2^z$, where $z > 2$ is a natural number, the first order box $\mathbf{P}_{n/m}$ with the recursive topology of the first type can be identically represented in the topology of the second type. Then, the $\mathbf{P}_{2n/2m+n}$ box (see Figure 4.2*a*) with the recursive topology of the first type can be identically represented in the topology of the second type. Indeed, on account of this assumption, each of the $\mathbf{P}_{n/m}$ boxes

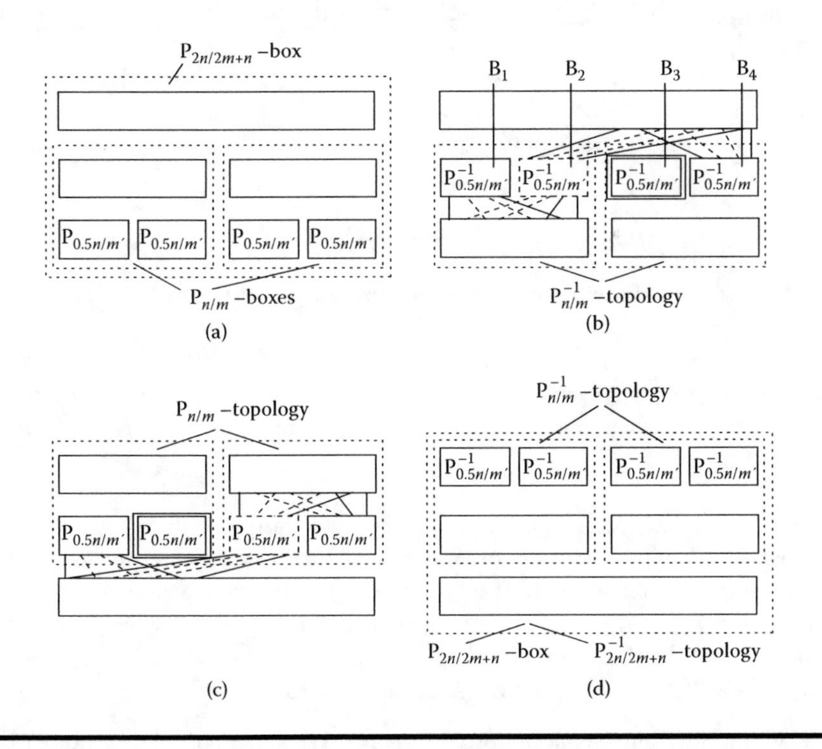

Figure 4.2 Topological transformations of the $P_{2n/2m+n}$ box: (*a*) initial representation in the recursive topology of the first type; (*b, c*) combined intermediate representations; (*d*) representation in recursive topology of the second type.

in Figure 4.2*a* can be represented in the $\mathbf{P}^{-1}_{n/m}$ topology, as is shown in Figure 4.2*b*. Transposing the inner $\mathbf{B}_2$ and $\mathbf{B}_3$ boxes having the $\mathbf{P}^{-1}_{0.5n/m'}$ topology (where $m' = 0.5(m - 0.5n)$), we transform two bottom layers $\mathbf{L}^{-1}_{(z)}$ into one linking layer $\mathbf{L}^{-1}_{(z+1)}$ of the second type topology and, simultaneously, divide the top linking layer $\mathbf{L}_{(z+1)}$ of the $\mathbf{P}_{2n/2m+n}$ topology into two top linking layers $\mathbf{L}_{(z)}$ of the $\mathbf{P}_{n/m}$ topology (see Figure 4.2*c*). Such transformations of the linking layers are provided as follows: Each $\mathbf{P}_{2/1}$ element of the right half of the $\mathbf{L}_{(z+1)}$ layer is connected with $\mathbf{B}_2$ and $\mathbf{B}_4$ boxes, and each $\mathbf{P}_{2/1}$ element of the left half of the $\mathbf{L}_{(z+1)}$ layer is connected with $\mathbf{B}_1$ and $\mathbf{B}_3$ boxes. Therefore, after the transposition of the $\mathbf{B}_2$ and $\mathbf{B}_3$ boxes, which retains the internal connections between all switching elements, the right (left) half of the top active layer has connection only with the right (left) half of the second active layer.

In Figure 4.2*c*, each of the $\mathbf{P}_{n/m}$ topologies can be transformed into the $\mathbf{P}^{-1}_{n/m}$ topology, as is shown in Figure 4.2*d*. Thus, we have got the identical representation of the $\mathbf{P}_{2n/2m+n}$ box in the topology of the second type, i.e., the given box can be represented in two mutually inverse topologies; therefore, for all V there exists V' such that $\Pi_{V'} = \Pi_V^{-1}$, where $\Pi_{V'}$ and Π_V are bit permutation modifications implemented with the given box $\mathbf{P}_{2n/2m+n}$ under the control of the controlling vectors V and V', respectively.

It is sufficiently evident that arbitrary CP box $\mathbf{P}^{-1}_{n/m}$ of the first order can be represented in the $\mathbf{P}_{n/m}$ topology, because arbitrary CP box $\mathbf{P}_{n/m}$ of the first order can be represented in the $\mathbf{P}^{-1}_{n/m}$ topology.

4.1.2 Topologies of the Higher Orders

In this subsection, we will give some generalization of Statement 4.2.

Statement 4.3. *Each of the* $\mathbf{P}_{n/m}$ *and* $\mathbf{P}^{-1}_{n/m}$ *boxes having the order* $h = 1, 2, \ldots, n/4, n$ *and constructed using recursive mechanisms of the first, second, and third types implements a set of fixed bit permutations that can be represented as a set of pairs of mutual inverses* Π_V *and* $\Pi_{V'} = \Pi_V^{-1}$.

First, note that recursive construction mechanism of the third type produces CP boxes $\mathbf{P}_{n/m}$ having maximum order n and mirror symmetry. In this case, Corollary 4.1 proves Statement 4.3. The $\mathbf{P}_{n/m}$ (or $\mathbf{P}^{-1}_{n/m}$) boxes having recursive topology and the order $h = 2, 4, \ldots, n/4$ are constructed combining the recursive mechanisms of the first and the third types (or the second and the third types). The $\mathbf{P}_{n/m}$ box of the order $h = 2^t \le n/4$, where $t \le z - 2$, can be constructed in the following way:

1. Using the recursive mechanism of the first type, compose the first order box $\mathbf{P}_{n'/m'}$, where $n' = n/h = 2^{z-t}$. This box contains $s' = \log_2 n' = \log_2(n/h)$ active layers.

2. Set counter $i = 1$.
3. Using the recursive mechanism of the third type and two $\mathbf{P}_{n'/m'}$ boxes, compose the 2^ith order box $\mathbf{P}_{2n'/2m'+2n'}$. Then designate $\mathbf{P}_{n'/m'} \leftarrow \mathbf{P}_{2n'/2m'+2n'}$.
4. If $i < t$, then increment $i \leftarrow i + 1$, and go to step 3; otherwise designate $\mathbf{P}_{n/m} \leftarrow \mathbf{P}_{n'/m'}$.
5. STOP.

This construction algorithm produces the $\mathbf{P}_{n/m}$ box of the order h. Indeed, it performs t steps of the third type recursive construction mechanism, and each of these steps doubles the order of the constructed box. Therefore, the constructed box $\mathbf{P}_{n/m}$ has the order $h = 2^t$. Accordingly, for the construction algorithm, the number of active layers in the $\mathbf{P}_{n/m}$ box is

$$s = s' + 2t = \log_2(n/h) + 2\log_2 h = \log_2 nh. \tag{4.4}$$

To construct a CP box $\mathbf{P}_{n/m}$ of the order $h = n/4$, it is necessary to use at least $s = 2\cdot\log_2 n - 2$ active layers. The maximum order box $\mathbf{P}_{n/m'}$ contains only $s = 2\cdot\log_2 n - 1$ active layers. This is explained by using the $\mathbf{P}_{2/1}$ elements (that have the order 2) at the first step of the algorithm while constructing the maximum order CP boxes.

If, at the first step of the construction algorithm, we use two $\mathbf{P}_{n'/m'}$ boxes for which Statement 4.3 is valid, then at step 3 of the algorithm we get the CP box for which also Statement 4.3 is valid. Thus, we have the following statement:

Statement 4.4. *A step of the recursive construction of the third type produces a CP box implementing a set of fixed bit permutations that can be represented as a set of pairs of mutual inverses* Π_V *and* $\Pi_{V'} = \Pi_V^{-1}$ *if each of the used two CP boxes of smaller size implements a set of fixed bit permutations that can be represented as a set of pairs of mutual inverses.*

The recursive construction step of the third type can be applied to link two equal CP boxes or two different CP boxes (see Figure 4.3). In both cases, Statement 4.4 holds. Figure 4.4 illustrates the proof of this statement. Let controlling vector V of the box **F** be represented as concatenation of the controlling vectors V_t, V_1, V_2, and V_b, corresponding to the top layer, to the inner boxes $\mathbf{F}_1$ and $\mathbf{F}_2$, and to the bottom layer, respectively, i.e., $V = (V_t, V_1, V_2, V_b)$. On account of the assumption for arbitrary values V_1 and V_2, there exist vectors V_1' and V_2' such that $\mathrm{F}_1^{(V_1')} = \left(\mathrm{F}_1^{(V_1)}\right)^{-1}$ and $\mathrm{F}_2^{(V_2')} = \left(\mathrm{F}_2^{(V_2)}\right)^{-1}$. Now it is easy to see that $\mathrm{F}^{(V')} = \left(\mathrm{F}^{(V)}\right)^{-1}$ for $V' = (V_t', V_1', V_2', V_b')$, where $V_t' = V_b$ and $V_b = V_t'$. Indeed, let us represent the **F** box in the following form:

$$\mathrm{F} = \mathrm{L}_{(z)} \circ \pi_{(z)} \circ \left(\mathrm{F}_1 \,||\, \mathrm{F}_2\right) \circ \pi_{(z)}^{-1} \circ \mathrm{L}_{(z)}^{-1} \tag{4.5}$$

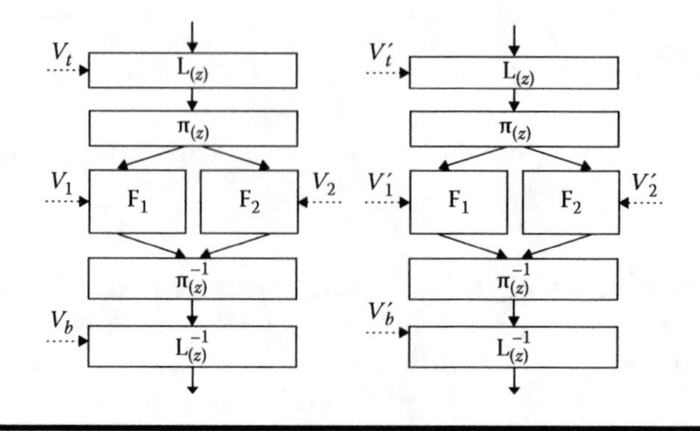

Figure 4.3 A step of recursive construction of the third type in arbitrary case of the CSPN based on minimum size controlled involutions.

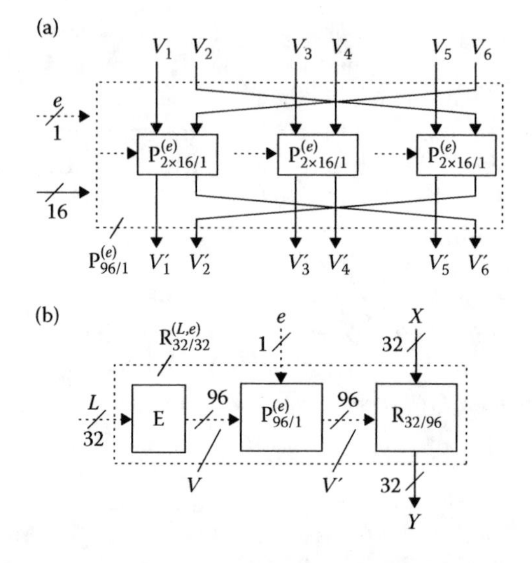

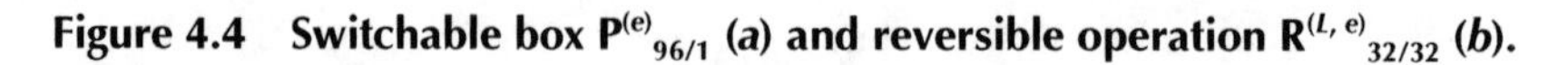

Figure 4.4 Switchable box $P^{(e)}_{96/1}$ (*a*) and reversible operation $R^{(L,e)}_{32/32}$ (*b*).

Then we can write:

$$\begin{aligned} F^{(V')} &= L_{(z)}^{(V'_t)} \circ \pi_{(z)} \circ \left(F_1^{(V'_1)} \,||\, F_2^{(V'_2)}\right) \circ \pi^{-1}_{(z)} \circ \left(L^{-1}_{(z)}\right)^{(V'_b)} \\ &= L_{(z)}^{(V_b)} \circ \pi_{(z)} \circ \left(F_1^{(V'_1)} \,||\, F_2^{(V'_2)}\right) \circ \pi^{-1}_{(z)} \circ \left(L^{-1}_{(z)}\right)^{(V_t)} \end{aligned} \tag{4.6}$$

$$F^{(V)} = L_{(z)}^{(V_t)} \circ \pi_{(z)} \circ \left(F_1^{(V_1)} \,||\, F_2^{(V_2)}\right) \circ \pi^{-1}_{(z)} \circ \left(L^{-1}_{(z)}\right)^{(V_b)} \tag{4.7}$$

$$F^{(V)} \circ F^{(V')} = L_{(z)}^{(V_t)} \circ \pi_{(z)} \circ \left(F_1^{(V_1)} \,||\, F_2^{(V_2)}\right) \circ \pi_{(z)}^{-1} \circ \left(L_{(z)}^{-1}\right)^{(V_b)} \circ L_{(z)}^{(V_b)} \circ$$

$$\circ \pi_{(z)} \circ \left(F_1^{(V_1')} \,||\, F_2^{(V_2')}\right) \circ \pi_{(z)}^{-1} \circ \left(L_{(z)}^{-1}\right)^{(V_t)}$$

$$= L_{(z)}^{(V_t)} \circ \pi_{(z)} \circ \left(F_1^{(V_1)} \,||\, F_2^{(V_2)}\right) \circ \pi_{(z)}^{-1} \circ \pi_{(z)} \circ \left(F_1^{(V_1')} \,||\, F_2^{(V_2')}\right) \circ \pi_{(z)}^{-1} \circ \left(L_{(z)}^{-1}\right)^{(V_t)}$$

$$= L_{(z)}^{(V_t)} \circ \pi_{(z)} \circ \left(F_1^{(V_1)} \,||\, F_2^{(V_2)}\right) \circ \left(F_1^{(V_1')} \,||\, F_2^{(V_2')}\right) \circ \pi_{(z)}^{-1} \circ \left(L_{(z)}^{-1}\right)^{(V_t)} \tag{4.8}$$

$$= L_{(z)}^{(V_t)} \circ \pi_{(z)} \circ \left(\left(F_1^{(V_1)} \circ F_1^{(V_1')}\right) \,||\, \left(F_2^{(V_2)} \circ F_2^{(V_2')}\right)\right) \circ \pi_{(z)}^{-1} \circ \left(L_{(z)}^{-1}\right)^{(V_t)}$$

$$= L_{(z)}^{(V_t)} \circ \pi_{(z)} \circ \pi_{(z)}^{-1} \circ \left(L_{(z)}^{-1}\right)^{(V_t)} = L_{(z)}^{(V_t)} \circ \left(L_{(z)}^{-1}\right)^{(V_t)} = O$$

Thus, the $F^{(V)} \circ F^{(V')}$ superposition represents the identical transformation **O**, i.e., $F^{(V')} = \left(F^{(V)}\right)^{-1}$. $F^{(V)}$ and $F^{(V')}$ are the Π_V and $\Pi_{V'}$ bit permutation modifications, i.e., we have proved that the CP box **F** implements a set of fixed bit permutations that can be represented as a set of pairs of mutual inverses Π_V and $\Pi_{V'} = \Pi_V^{-1}$. Statements 4.1 to 4.4 can be easily extended on the case of homogeneous (uniform) controlled substitution–permutation networks (CSPNs) constructed using controlled elementary involutions $\mathbf{Q}_{2/1}$, $\mathbf{R}_{2/1}$, and $\mathbf{Z}_{2/1}$. (Notion of the order of the CSPN is introduced with Definition 3.1 in Chapter 3, page 83.)

Thus, we have a number of CP and CSPN boxes, each of them implementing the pairs of mutually inverse transformation modifications. Therefore, the same box from this class of boxes can be used to implement direct and inverse controlled transformations. But for practical use of this property, we need some constructions that provide efficient implementation of this idea. In the next paragraphs, some approaches providing efficient constructions of switchable controlled operations (SCOs) are considered. The cryptographic application of such primitives relates to their use in the form of switchable data-dependent operations (SDDOs).

4.2 Reversible DDO Boxes

As mentioned earlier, Statements 4.1 to 4.4 can be formulated in a more general case that relates to uniform CSPNs constructed using the elementary controlled involutions $\mathbf{Q}_{2/1}$, $\mathbf{R}_{2/1}$, $\mathbf{Z}_{2/1}$, and $\mathbf{F}_{2/2}$ as standard building blocks or to nonuniform symmetric CSPNs. For example consider the following statement:

Statement 4.1′. *Suppose* $\mathbf{VT}^{(V)}$ *and* $(\mathbf{VT}^{-1})^{(V)}$ *are mutually inverse CSPN boxes having mirror symmetry topology. Then, each fixed transformation modification* Ψ_V *implemented with the* **VT** *box can be implemented also with the* $\mathbf{VT}^{-1}$ *box as fixed transformation modification* $\Psi'_{V'}$.

In this statement, we use notation **VT** (variable transformation) as general representation of the boxes $\mathbf{Q}_{n/m}$, $\mathbf{R}_{n/m}$, $\mathbf{Z}_{n/m}$, $\mathbf{U}_{n/m}$, and $\mathbf{\Phi}_{n/m'}$. Using the mirror symmetry topologies of the CSPNs, we can easily prove this statement. Indeed, for $V = (V_1, V_2, \ldots, V_s)$, one can select $V' = (V_s, V_{s-1}, \ldots, V_1)$, which proves the statement. Using the symmetric topologies presented in Table 3.10 of Chapter 3 (see page 87), we can construct the following rows of symmetric DDO boxes:

- $\mathbf{Q}_{2/1} \rightarrow \mathbf{Q}_{4/6} \rightarrow \mathbf{Q}_{8/20} \rightarrow \mathbf{Q}_{16/56} \rightarrow \mathbf{Q}_{32/144} \rightarrow \mathbf{Q}_{64/352} \rightarrow \mathbf{Q}_{128/832}$ (h has maximum value)
- $\mathbf{R}_{16/32}$ ($h = 1$) $\rightarrow \mathbf{R}_{32/96}$ ($h = 2$) $\rightarrow \mathbf{R}_{64/256}$ ($h = 4$) $\rightarrow \mathbf{R}_{128/640}$ ($h = 8$)
- $\mathbf{\Phi}_{4/8}$ ($h = 1$) $\rightarrow \mathbf{\Phi}_{8/32}$ ($h = 2$) $\rightarrow \mathbf{\Phi}_{16/96}$ ($h = 4$) $\rightarrow \mathbf{\Phi}_{32/256}$ ($h = 8$) $\rightarrow \mathbf{\Phi}_{64/640}$ ($h = 16$)

Corollary 4.1′. *An arbitrary CSPN having the mirror symmetry topology implements a set of fixed transformation modifications that can be represented as a set of pairs of mutual inverses* Ψ_V *and* $\Psi_{V'} = \Psi_V^{-1}$.

Statement 4.3′. *Each of the boxes* $\mathbf{VT}_{n/m}$ *and* $\mathbf{VT}_{n/m}^{-1}$ *having the order* $h = 1, 2, \ldots, n/4, n$ *and constructed using recursive mechanisms of the first, second, and third types implements a set of fixed transformation modifications that can be represented as a set of pairs of mutual inverses* Ψ_V *and* $\Psi_{V'} = \Psi_V^{-1}$.

A corollary from Statement 4.3′ is the following: *two mutually inverse DDO boxes* $\mathbf{VT}^{(V)}$ *and* $(\mathbf{VT}^{-1})^{(V)}$ *of the orders* $h = 1, 2, \ldots, n/4, n$*, which are constructed using controlled elements (CEs) that are involutions and the recursive mechanisms of the first, second, and third types, implement the same set of the fixed transformation modifications* $\{\Psi_V\} = \{\Psi_V^{-1}\}$.

At this point, we will consider a mechanism providing for the construction of the nonuniform, nonsymmetric CSPNs for which Statement 4.3′ is also valid (naturally, the same mechanism provides also for the construction of the uniform nonsymmetric CSPNs for which Statement 4.3′ is valid). Statement 4.3′ reflects some fundamental properties of the $\mathbf{P}_{n/m}$, $\mathbf{F}_{n/m}$, and $\mathbf{\Phi}_{n/m'}$ boxes that represent interest for constructing switchable DDO (SDDO) boxes that are very attractive as cryptographic primitive. This statement shows that, in general, for each value V, we can put into correspondence another value V' such that $\mathbf{VT}^{(V')} = (\mathbf{VT}^{(V)})^{-1}$. Thus, the **VT** box modifications corresponding to values V and V' are mutual inverses. If we design an extension box that is able to generate the V' value corresponding to arbitrary value V, then we will have a switchable **VT** box. Such SDDOs can perform mutually inverse modifications $\mathbf{VT}^{(V)}$ and $\mathbf{VT}^{(V')} = (\mathbf{VT}^{(V)})^{-1}$, corresponding to arbitrary given value V.

Let us introduce the notion of the switchable extension box $\mathbf{E}^{(e)}(X)$, the output of which depends on the input vector X and the controlling bit e (which is called switching bit), i.e., we have $V = V(e) = V^{(e)}$. If a DDO box is controlled with the output of the switchable extension box $\mathbf{E}^{(e)}$, then we have an SDDO box $\mathbf{VT}^{(V,e)}$ that includes the $\mathbf{E}^{(e)}$ box as its internal part. An SDDO box is reversible if, for all values X, the following condition is satisfied:

$$\mathbf{VT}^{(V(0))} = (\mathbf{VT}^{(V(1))})^{-1} \text{ and } \mathbf{VT}^{(V(1))} = (\mathbf{VT}^{(V(0))})^{-1} \tag{4.9}$$

We will call the reversible RSDDOs simply SDDOs or reversible DDOs (RDDOs). The formal presentation of the reversible controlled operations uses the following definitions:

Definition 4.1. *Let* $\{\mathbf{F}_0, \mathbf{F}_1, \ldots, \mathbf{F}_{2m-1}\}$ *be some set of the single-type operations defined by the formula* $Y = \mathbf{F}_i = \mathbf{F}_i(X)$, where $i = 1, 2, \ldots, 2^{m-1}$, *$X$ is an n-bit input binary vector, and Y is the output. Then, the V-dependent operation* $\mathbf{F}^{(V)}$ *defined by the formula* $Y = \mathbf{F}^{(V)}(X) = \mathbf{F}_V(X)$, *where V is an m-bit controlling vector, is called the controlled operation. The operations* $\mathbf{F}_1, \mathbf{F}_2, \ldots, \mathbf{F}_{2m}$ *are called modifications of the controlled operation* $\mathbf{F}^{(V)}$.

Definition 4.2. *Let* $\{\mathbf{F}_0, \mathbf{F}_1, \ldots, \mathbf{F}_{2m-1}\}$ *be the set of the modifications of the controlled operation* $\mathbf{F}^{(V)}$. *The controlled operation* $(\mathbf{F}^{-1})^{(V)}$ *implementing the set of modifications* $\{(\mathbf{F}^{-1})_1, (\mathbf{F}^{-1})_2, \ldots, (\mathbf{F}^{-1})_{2m-1}\}$ *is called inverse of* $\mathbf{F}^{(V)}$ *if, for all values V, we have* $(\mathbf{F}^{-1})_V = \mathbf{F}_V^{-1}$, *i.e., the modifications* $\mathbf{F}_V$ *and* $(\mathbf{F}^{-1})_V$ *are mutual inverses for all* $V \in \{1, 2, \ldots, 2^m - 1\}$.

Definition 4.3. *Let* $\mathbf{\Omega}^{(e)}$ *(where* $e = \{0, 1\}$*) be some e-dependent operation containing two modifications* $\mathbf{\Omega}_1 = \mathbf{\Omega}^{(0)}$ and $\mathbf{\Omega}_2 = \mathbf{\Omega}^{(1)}$, where $\mathbf{\Omega}_2 = \mathbf{\Omega}_1^{-1}$. *Then, the operation* $\mathbf{\Psi}^{(e)}$ *is called reversible.*

Definition 4.4. *Let two modifications of the reversible operation* $\mathbf{\Omega}^{(e)}$ *be mutual inverses* $\mathbf{\Omega}^{(0)} = \mathbf{F}^{(V)}$ *and* $\mathbf{\Omega}^{(1)} = (\mathbf{F}^{-1})^{(V)}$. *Then, the operation* $\mathbf{\Omega}^{(e)}$ *is called reversible controlled operation* $\mathbf{F}^{(V,e)}$ *for which we have* $\mathbf{\Omega}^{(0)} = \mathbf{F}^{(V,0)}$ *and* $\mathbf{\Omega}^{(1)} = \mathbf{F}^{(V,1)}$.

A great variety of different reversible controlled operations can be implemented on the basis of CSPNs. In the following text, we consider several design approaches and show that the most efficient variants correspond to the use of mirror symmetry CSPNs. Reversible controlled operations (RCOs) of the orders $h = 1, 2, \ldots, n$ can also be easily constructed. In this section, some designs of RSDDO boxes, which are efficient for hardware implementation, are considered. The following designs are described:

- Symmetric topology of the CSPN and general distribution of the controlling bits
- Symmetric topology of the CSPN and symmetric distribution of the controlling bits corresponding to different halves of the controlling data subblock
- Nonsymmetric topology of the CSPN and nonsymmetric distribution of the controlling bits
- SDDO boxes using inversion of the controlling bits as switching mechanism

4.2.1 SDDO Boxes with Symmetric Topology

The proof of Statement 4.1′ presented indicates a direct method of constructing the switchable CSPN having mirror symmetry topology. Let us consider the case of the mirror symmetry topology, for example, of the $\mathbf{R}_{32/96}$ box of the second order. Due to the symmetric structure of $\mathbf{R}_{32/96}$, its modifications $\mathbf{R}^{(V)}_{32/96}$, where $V = (V_1, V_2, \ldots, V_6)$, and $\mathbf{R}^{(V')}_{32/96}$, where $V' = (V_6, V_5, \ldots, V_1)$, are mutually inverse. The reversible $\mathbf{R}^{(V,e)}_{32/96}$ box can be implemented using the very simple transposition box $\mathbf{P}^{(e)}_{96/1}$ that represents some single-layer CP box comprising three parallel single-layer boxes $\mathbf{P}^{(e)}_{2\times16/1}$ (Figure 4.4*a*). The input of each $\mathbf{P}^{(e)}_{2\times16/1}$ box is divided into 16-bit left and 16-bit right inputs. The box $\mathbf{P}^{(e)}_{2\times16/1}$ represents a cascade of 16 switching elements $\mathbf{P}^{(e)}_{2/1}$ controlled with the same bit e. For example, $\mathbf{P}^{(0)}_{2\times16/1}(U) = U$ and $\mathbf{P}^{(1)}_{2\times16/1}(U) = U' = (U_h, U_l)$, where $U = (U_l, U_h) \in \{0, 1\}^{32}$. The left (right) inputs of the $\mathbf{P}^{(e)}_{2/1}$ boxes correspond to the left (right) 16-bit input of the box $\mathbf{P}^{(e)}_{2\times16/1}$. If the input vector of the box $\mathbf{P}^{(e)}_{96/1}$ is $(V_1, V_2, \ldots, V_6)$, then, at the output of $\mathbf{P}^{(e)}_{96/1}$, we have $V' = (V_1, V_2, \ldots, V_6)$ if $e = 0$, or $V' = (V_6, V_5, \ldots, V_1)$ if $e = 1$. The structure of the RCO box $\mathbf{R}^{(L,e)}_{32/32}$ is shown in Figure 4.4*b*.

By analogy, one can construct the following variants of reversible controlled operations: $\mathbf{Z}^{(L,e)}_{32/32}$, $\mathbf{Q}^{(L,e)}_{32/32}$, $\mathbf{\Phi}^{(L,e)}_{32/32}$, $\mathbf{U}^{(L,e)}_{64/64}$, $\mathbf{R}^{(L,e)}_{64/64}$, $\mathbf{Q}^{(L,e)}_{64/64}$, and $\mathbf{\Phi}^{(L,e)}_{64/64}$ that correspond to using the $\mathbf{Z}^{(V,e)}_{32/96}$, $\mathbf{Q}^{(V,e)}_{32/96}$, $\mathbf{\Phi}^{(V,e)}_{32/192}$, $\mathbf{U}^{(V,e)}_{64/192}$, $\mathbf{R}^{(V,e)}_{64/192}$, $\mathbf{Q}^{(V,e)}_{64/192}$, and $\mathbf{\Phi}^{(V,e)}_{64/384}$ switchable CSPNs, respectively. Using these reversible operations instead of the switchable data-driven primitive (DDP) boxes in the frame of cryptoschemes of the firmware-suitable ciphers Cobra-F64a, Cobra-F64b, and Cobra-S128 [48], one can develop a lot of fast and hardware-efficient ciphers with 64-, 128-, and 256-bit input data blocks. The updated versions of these algorithms provide secure encryption with less number of rounds; therefore, they possess higher performance. A novel feature of reversible operations such as cryptographic primitive is the possibility to eliminate the weak keys in iterative ciphers with simple key scheduling [41] as well as to eliminate the appearance of periodicity in the iterative encryption procedures using the fixed round key [41, 45]. The last fact is efficient enough to thwart the so-called slide attacks against encryption algorithms with very simple key scheduling [7].

When the SDDO boxes are used in block ciphers, the transformed data subblock and controlling data subblock L have the same size. In such cases, SDDO boxes can be denoted as $\mathbf{F}^{(L,e)}_{n/n}$ or simply $\mathbf{F}^{(L,e)}_{n}$. In such designation we assume that the extension boxes are incorporated in the $\mathbf{F}_{n/n}^{(L,e)}$ boxes as internal parts. However, in the simplified designation of the RCOs, there is no indication of the number of active layers incorporated in the CSPNs used as a core unit of the RCOs. In the following text, we will use both variants for designating different variants of the RCO boxes.

4.2.2 Hardware Efficient SDDOs

The RCO design already considered imposes no restrictions on the distribution of the controlling bits; however, in the DDO-based ciphers, the m-bit controlling vector V usually depends on some n-bit controlling data subblock L. The V vector is formed as an output of the extension operation $\mathbf{E}$ performed on L. Because we have $m = 3n$, one can propose the RCO construction with reduced implementation cost, designing respective mechanism for swapping some components of L. For example, one can use the following design in which the L_l and L_h components of the data subblock $L = (L_l, L_h)$ are extended using two symmetric extension boxes $\mathbf{E}_1$ and $\mathbf{E}_2$. Let the outputs of $\mathbf{E}_1$ and $\mathbf{E}_2$ be $V' = (V_1, V_2, V_3)$ and $V'' = (V_4, V_5, V_6)$, respectively. Thus, the $\mathbf{E}_1$ and $\mathbf{E}_2$ boxes can represent a single extension box $\mathbf{E}$ with symmetric structure, which produces the controlling vector $V = (V', V'')$, corresponding to the symmetric distribution of the bits of the $L_l = (l_1, \ldots, l_{n/2})$ and $L_h = (l_{n/2}, l_{n/2+1}, \ldots, l_n)$ components of the controlling data subblock. In this case, swapping the L_l and L_h components means swapping components V_j and V_{7-j} for all $j \in \{1, 2, \ldots, 6\}$. Considering the symmetric extension box $\mathbf{E}$ as its internal part, the RCO box can be denoted as some box $\mathbf{F}_{n/n}^{(L,e)}$ with n-bit controlling input. For the $\mathbf{F}_{32/32}^{(L,e)}$ or $\mathbf{F}_{64/64}^{(L,e)}$ operations, we can propose the $\mathbf{E}$ box having the following structure:

$$V_1 = L_l;\ V_2 = L_l^{>>>6};\ V_3 = L_l^{>>>12};\ V_4 = L_h^{>>>12};\ V_5 = L_h^{>>>6};\ V_6 = L_h. \qquad (4.10)$$

For the $\mathbf{\Phi}_{32/32}^{(L,e)}$ and $\mathbf{\Phi}_{64/64}^{(L,e)}$ operations (that include the switchable CSPNs $\mathbf{\Phi}_{32/192}^{(V,e)}$ and $\mathbf{\Phi}_{64/384}^{(V,e)}$, correspondingly), one can use the $\mathbf{E}$ box described as follows:

$$V_1 = L_l;\ V_2 = L_l^{>>>10};\ V_3 = L_l^{>>>20};\ V_4 = L_h^{>>>20};\ V_5 = L_h^{>>>10};\ V_6 = L_h. \qquad (4.11a)$$

$$Z_1 = L_l^{>>>5};\ Z_2 = L_l^{>>>15};\ Z_3 = L_l^{>>>25};\ Z_4 = L_h^{>>>25};\ Z_5 = L_h^{>>>15};\ Z_6 = L_h^{>>>5} \qquad (4.11b)$$

In the aforesaid case, symmetric extension boxes $\mathbf{E}_1$ and $\mathbf{E}_2$ form the outputs $(V', Z') = (V_1, Z_1, V_2, Z_2, V_3, Z_3)$ and $(V'', Z'') = (V_4, Z_4, V_5, Z_5, V_6, Z_6)$, respectively.

This hardware-efficient design of the RCO boxes is illustrated in Figure 4.5 for the $\mathbf{F}_{32/32}^{(L,e)}$ (a) and $\mathbf{F}_{64/64}^{(L,e)}$ (b) boxes. To embed the reversibility mechanism in the $\mathbf{F}_{32/32}^{(L,e)}$ (*a*) and $\mathbf{F}_{64/64}^{(L,e)}$ (*b*) boxes, we need only 96 and 192 extra NAND gates, correspondingly, against 288 and 576 extra NAND gates for the design as in Figure 4.4.

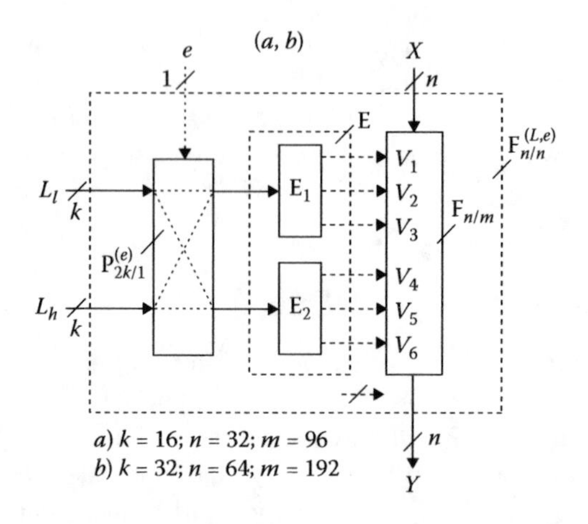

Figure 4.5 Reversible $F^{(L,e)}_{32/32}$ (*a*) and $F^{(L,e)}_{64/64}$ (*b*) boxes.

Higher efficiency of the hardware implementation is due to the symmetric distribution of the controlling bits corresponding to two halves of the controlling data subblock.

4.2.3 General Design of the SDDO Boxes of Different Orders

Taking into account symmetric topologies of the CSPNs described in Chapter 3, Section 3.3, we see that the design presented in Subsections 4.2.1 and 4.2.2 allows us to construct the SDDO boxes with different values of the order h and input size n; but this does not allow for all possible combinations of the h and n values corresponding to the minimum values of the active layers for the given h and n values that can be represented as $h = 2^t$ and $n = 2^z$ for some natural numbers t and z (these cases are the most interesting for practice). The SDDO designs based on the symmetric topologies cover the cases of the even number of active layers $s = \log_2 nh$. In this section, we describe another design of the SDDOs, which provides the construction for arbitrary values h and n independently of the oddness of the value $\log_2 nh$.

A more general method to synthesize the SDDOs is shown in Figure 4.6 where $\mathbf{F}_{n/m}$ and $\mathbf{F}^{-1}_{n/m}$ are the DDO boxes of the order h that are linked with the linking fixed permutations $\pi_{(z+1)}$ and $\pi'_{(z+1)}$. The $\pi'_{(z+1)}$ permutation represents the superposition of two fixed permutations, i.e., $\pi'_{(z+1)} = \mathbf{T}^{\circ}\pi^{-1}_{(z+1)}$ where $\mathbf{T}$ is the transposition operation that swaps the outputs of the $\mathbf{F}_{n/m}$ and $\mathbf{F}^{-1}_{n/m}$ boxes. The linking permutations $\pi_{(z+1)}$ and $\pi^{-1}_{(z+1)}$ are described in Subsection 4.1.1. This construction mechanism resembles the recursive construction of the third type, except that the linked boxes are different and their outputs are transposed. Besides, the $\mathbf{F}_{n/m}$ and $\mathbf{F}^{-1}_{n/m}$ boxes can be constructed using mutually inverse CEs $\mathbf{F}_{2/1}$ and $\mathbf{F}^{-1}_{2/1}$ that are not involutions.

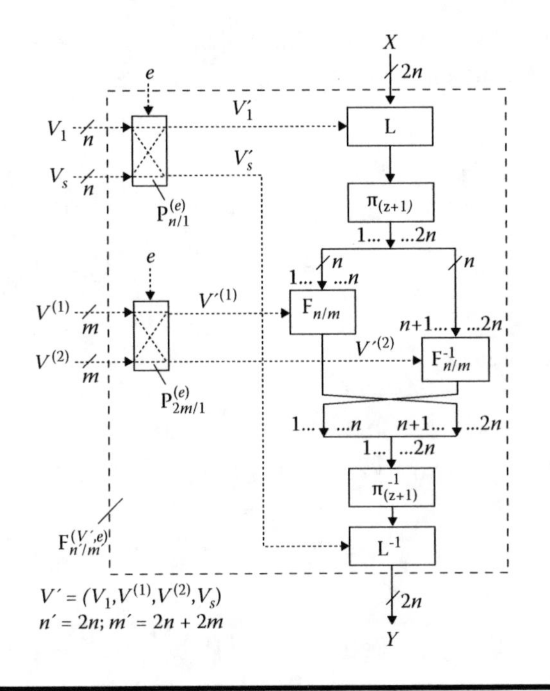

Figure 4.6 Construction of the switchable DDOs of different orders.

Thus, the considered mechanism allows us to construct the SDDOs that are nonsymmetric and nonuniform. An SDDO box $\mathbf{F}_{n'/m'}$ of the order $h' = 2^{t'} \le n'/4$, where $t' \le z' - 2$, can be constructed in the following way:

1. Construct the hth order box $\mathbf{F}_{n/m}$, where $h = h'/2$ and $n = n'/2 = 2^{z'-1}$, using the recursive construction mechanisms of the first and the third types. This box contains $s = \log_2 nh = \log_2 nh' - 1$ active layers.
2. Construct the hth order box $\mathbf{F}^{-1}_{n/m}$, where $h = h'/2$ and $n = n'/2 = 2^{z'-1}$, using the recursive construction mechanism of the second and the third types. This box contains $s = \log_2 nh = \log_2 nh' - 1$ active layers.
3. Using the mechanism presented in Figure 4.6, construct the SDDO box $\mathbf{F}_{n'/m'} = \mathbf{F}_{2n/2m+2n}$ of the order $h' = 2h$. This box contains $s' = s + 2 = \log_2 nh' + 1 = \log_2 2nh' = \log_2 n'h'$ active layers.

The first two steps can be performed with the algorithm presented in subsection 4.1.2. Thus, the SDDO boxes can be constructed using several steps of the recursive construction of the first, second, and third types and one step defined in Figure 4.6. The last step multiplies by two the order h of the united boxes $\mathbf{F}_{n/m}$ and $\mathbf{F}^{-1}_{n/m}$, similar to the recursive construction step of the third type. Therefore, the constructed SDDO box possesses the minimum number of active layers for given values n' and h'.

The proof that the mechanism specified in Figure 4.6 provides the correct construction of the SDDO box is the following:

We have

$$\left(F_{n/m} \,||\, F_{n/m}^{-1}\right) \circ T = T \circ \left(F_{n/m}^{-1} \,||\, F_{n/m}\right),\ V_1' = V_1,\ V'^{(1)} = V^{(1)},\ V'^{(2)} = V^{(2)},\ V_s' = V_s,$$
$$\text{if } e = 0, \text{ and } V_1' = V_s,\ V'^{(1)} = V^{(2)},\ V'^{(2)} = V^{(1)},\ V_s' = V_1, \text{ if } e = 1.$$

Therefore we can write:

$$F_{n'/m'}^{(V',0)} \circ F_{n'/m'}^{(V',1)} = L^{(V_1)} \circ \pi_{(z+1)} \circ \left(F_{n/m}^{(V'^{(1)})} \,||\, (F_{n/m}^{-1})^{(V'^{(2)})}\right) \circ T \circ \pi_{(z+1)}^{-1} \circ \left(L^{-1}\right)^{(V_s)} \circ$$
$$\circ L^{(V_s)} \circ \pi_{(z+1)} \circ \left(F_{n/m}^{(V'^{(2)})} \,||\, (F_{n/m}^{-1})^{(V'^{(1)})}\right) \circ T \circ \pi_{(z+1)}^{-1} \circ \left(L^{-1}\right)^{(V_1)} =$$
$$= L^{(V_1)} \circ \pi_{(z+1)} \circ T \circ \left((F_{n/m}^{-1})^{(V'^{(2)})} \,||\, F_{n/m}^{(V'^{(1)})}\right) \circ \pi_{(z+1)}^{-1} \circ$$
$$\circ \pi_{(z+1)} \circ \left(F_{n/m}^{(V'^{(2)})} \,||\, (F_{n/m}^{-1})^{(V'^{(1)})}\right) \circ T \circ \pi_{(z+1)}^{-1} \circ \left(L^{-1}\right)^{(V_1)} =$$
$$= L^{(V_1)} \circ \pi_{(z+1)} \circ T \circ \left((F_{n/m}^{-1})^{(V'^{(2)})} \,||\, F_{n/m}^{(V'^{(1)})}\right)$$
$$\circ \left(F_{n/m}^{(V'^{(2)})} \,||\, (F_{n/m}^{-1})^{(V'^{(1)})}\right) \circ T \circ \pi_{(z+1)}^{-1} \circ \left(L^{-1}\right)^{(V_1)} =$$
$$= L^{(V_1)} \circ \pi_{(z+1)} \circ T \circ T \circ \pi_{(z+1)}^{-1} \circ \left(L^{-1}\right)^{(V_1)} = L^{(V_1)} \circ \left(L^{-1}\right)^{(V_1)} = \mathbf{O} \quad (4.12)$$

where $\mathbf{O}$ is the identical transformation. Thus, the $F_{n'/m'}^{(V',0)} \circ F_{n'/m'}^{(V',1)}$ superposition represents the identical transformation, i.e., $F_{n'/m'}^{(V',1)} = \left(F_{n'/m'}^{(V',0)}\right)^{-1}$.

With the use of considered mechanism, nonsymmetric SDDO boxes of the maximum order (i.e., $h' = n'$) can be constructed using $s' = 2 \cdot \log_2 n'h' - 1$ active layers. Such boxes contain odd number of active layers. Using the recursive construction of the third type, we can construct the symmetric DDO boxes of the maximum order, which represent the only example of symmetric DDO boxes with odd number of active layers (we consider the topologies with minimum number of layers for given values n and h). Therefore, SDDO boxes of maximum order can also be constructed using the methods presented in Sections 4.2.1 and 4.2.2.

4.2.4 The RCO Design Based on CE with Mutual Inverse Modifications

In the designs of SDDOs, the method providing for the reversibility of the SDDOs is the change of the controlling bits distribution. There is another efficient method for constructing SDDOs, which relates to specific cases of the CEs used as standard building blocks of the CSPNs, that represent the core part of the RCO boxes. In the second method, inverting of all of the controlling bits is used to reverse the transformations performed with the CSPNs. This method also provides for construction of the RCO boxes that are efficient for cheap hardware implementation. It is based on the use of CEs that implement mutually inverse modifications $\mathbf{F}^{(0)}_{2/1}$ and $\mathbf{F}^{(1)}_{2/1}$ for which we have $\mathbf{F}^{(0)}_{2/1} = (\mathbf{F}^{(1)}_{2/1})^{-1}$. In this method, the symmetric topology of the CSPNs combined with the symmetric distribution of the controlling bits is used (now we consider the symmetric distribution of the controlling bits corresponding to Definition 4.5). The symmetric distribution of the controlling bits is provided by the **E** extension box with m-bit output, which is constructed as shown in Figure 4.7 on the basis of arbitrary extension box $\mathbf{E}_3$ with $(m/2)$-bit output. Figures 4.7*a* and *b*, where $E = \{e\}^{2k}$ is the concatenation of $2k$ bits, all of which are equal to the switching bit e, illustrates the construction of the RCO boxes $\mathbf{Q}^{(V,e)}_{32/32}$, (*a*) and $\mathbf{Q}^{(V,e)}_{64/64}$ (*b*), where the $\mathbf{Q}^{(V)}_{32/128}$ (*a*) and $\mathbf{Q}^{(V)}_{64/256}$ (*b*) boxes have symmetric topologies (see Table 3.10 in Chapter 3). It is easy to show that this construction provides the relation

$$\mathbf{Q}^{(V,0)}_{n/n} = (\mathbf{Q}^{(V,1)}_{n/n})^{-1}. \tag{4.13}$$

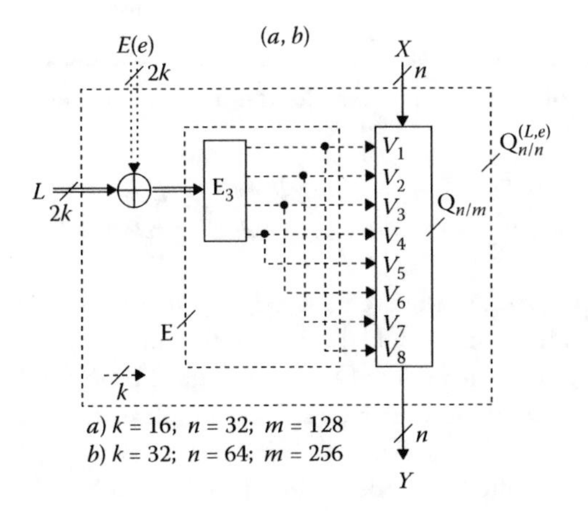

Figure 4.7 Reversible $\mathbf{Q}^{(L,e)}_{32/32}$ (*a*) and $\mathbf{Q}^{(L,e)}_{64/64}$ (*b*) boxes.

Table 4.1 Nonlinear CEs with mutual inverse modifications $\mathbf{F}^{(0)}_{2/1}$ and $\mathbf{F}^{(1)}_{2/1}$

CEs	Type of CEs	$y_1 = f_1(x_1, x_2, v)$	$y_2 = f_2(x_1, x_2, v)$	$y_1 \oplus y_2 = f_3(x_1, x_2, v)$
(q,u)	$\mathbf{Q}_{2/1}$	$vx_1 \oplus x_2$	$vx_2 \oplus x_1 \oplus x_2$	$vx_1 \oplus vx_2 \oplus x_1$
(u,q)	$\mathbf{Q}_{2/1}$	$vx_1 \oplus x_1 \oplus x_2$	$vx_2 \oplus x_1$	$vx_1 \oplus vx_2 \oplus x_2$
(s,v)	$\mathbf{Q}_{2/1}$	$vx_1 \oplus x_2 \oplus v$	$vx_2 \oplus x_1 \oplus x_2 \oplus v \oplus 1$	$vx_1 \oplus vx_2 \oplus x_1 \oplus 1$
(v,s)	$\mathbf{Q}_{2/1}$	$vx_1 \oplus x_1 \oplus x_2 \oplus v \oplus 1$	$vx_2 \oplus x_1 \oplus v$	$vx_1 \oplus vx_2 \oplus x_2 \oplus 1$
(t,w)	$\mathbf{Q}_{2/1}$	$vx_1 \oplus x_2 \oplus 1$	$vx_2 \oplus v \oplus x_1 \oplus x_2$	$vx_1 \oplus vx_2 \oplus x_1 \oplus v \oplus 1$
(w,t)	$\mathbf{Q}_{2/1}$	$vx_1 \oplus x_1 \oplus x_2 \oplus 1$	$vx_2 \oplus x_1 \oplus v \oplus 1$	$vx_1 \oplus vx_2 \oplus x_2 \oplus v$
(r,x)	$\mathbf{Q}_{2/1}$	$vx_1 \oplus x_2 \oplus v \oplus 1$	$vx_2 \oplus x_1 \oplus x_2 \oplus 1$	$vx_1 \oplus vx_2 \oplus x_1 \oplus v$
(x,r)	$\mathbf{Q}_{2/1}$	$vx_1 \oplus x_1 \oplus x_2 \oplus v$	$vx_2 \oplus x_1 \oplus 1$	$vx_1 \oplus vx_2 \oplus v \oplus x_2 \oplus 1$

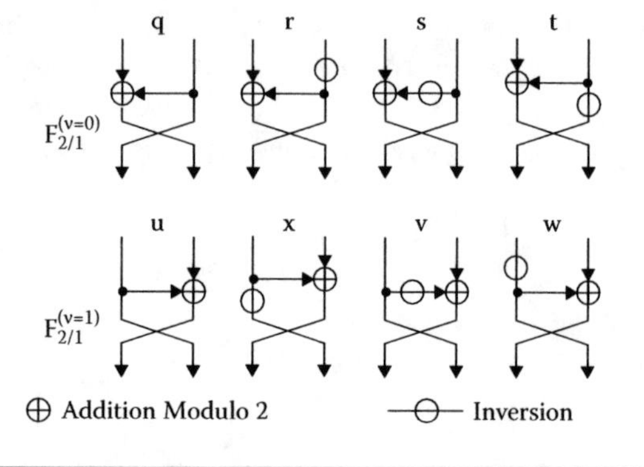

Figure 4.8 Pairs of mutually inverse elementary transformations of the type $(x_1, x_2) \rightarrow (y_1, y_2)$.

Definition 4.5. *Distribution of the controlling bits of the $V = (V_1, \ldots, V_{s/2})$ vector is called symmetric if for all $j = 1, \ldots, [s/2]$, the following relation holds: $V_j = V_{s-j+1}$.*

This definition relates to the even and odd number of active layers s.

Table 4.1 lists all nonlinear CEs having mutual inverse modifications $\mathbf{F}^{(0)}_{2/1}$ and $\mathbf{F}^{(1)}_{2/1}$. These CEs are attributed to the $\mathbf{Q}_{2/1}$ type of the CEs. Note that Table 4.1 actually lists four pairs of mutual inverse CEs. For example, the (q,u) and (u,q) elements are mutual inverses. Figure 4.8 presents typical examples of the CEs $\mathbf{F}_{2/1}$ that can be used in the considered method of designing the RCO.

Other designs of the RCO based on the switching mechanism comprising inverting the controlling bits are possible while using the $\mathbf{F}_{2/2}$ elements that implement the elementary-transformation modifications satisfying the following conditions:

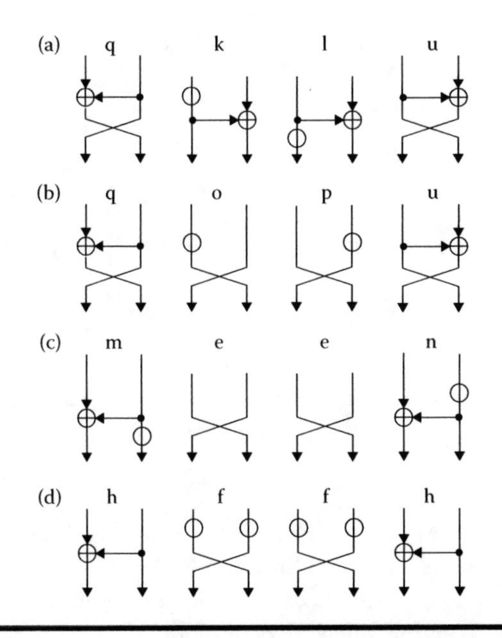

Figure 4.9 The $F_{2/2}$ elements implementing two pairs of mutually inverse elementary transformations: (*a*) element (q,k,l,u); (*b*) element (q,o,p,u); (*c*) element (m,e,e,n); (*d*) element (h,f,f,h).

$$\mathbf{F}^{(11)} = (\mathbf{F}^{(00)})^{-1} \text{ and } \mathbf{F}^{(10)} = (\mathbf{F}^{(01)})^{-1} \tag{4.14}$$

There are a sufficiently large number of different nonlinear CEs $\mathbf{F}_{2/2}$ possessing such property. They can be divided into the following subsets:

- CEs implementing four different elementary transformations (Figures 4.9*a* and *b* presents two examples of such CEs.)
- CEs implementing three different elementary transformations (see Figure 4.9*c*)
- CEs implementing two different elementary transformations (see Figure 4.9*d*)

The outputs of the element (q,k,l,u) shown in Figure 4.9*a* are described by the following nonlinear BFs:

$$y_1 = vzx_1 \oplus vx_2 \oplus zx_2 \oplus zx_1 \oplus vx_1 \oplus v \oplus z \oplus x_2 \tag{4.15}$$

$$y_2 = vzx_2 \oplus vz \oplus x_1 \oplus x_2 \oplus z \tag{4.16}$$

$$y_1 \oplus y_2 = vzx_1 \oplus vzx_2 \oplus vz \oplus vx_1 \oplus vx_2 \oplus zx_1 \oplus zx_2 \oplus v \oplus x_1 \tag{4.17}$$

The outputs of the element (m,e,e,n) shown in Figure 4.9*c* are described as follows:

$$y_1 = vx_1 \oplus zx_1 \oplus vz \oplus x_1 \oplus x_2 \quad (4.18)$$

$$y_2 = vx_1 \oplus zx_1 \oplus vx_2 \oplus zx_2 \oplus x_2 \oplus v \oplus z \oplus 1 \quad (4.19)$$

$$y_1 \oplus y_2 = vx_2 \oplus zx_2 \oplus vz \oplus x_1 \oplus v \oplus z \oplus 1 \quad (4.20)$$

The outputs of the element (h,f,f,h) shown in Figure 4.9*d* represent the following BFs:

$$y_1 = vx_1 \oplus zx_1 \oplus x_1 \oplus x_2 \oplus v \oplus z \quad (4.21)$$

$$y_2 = vx_1 \oplus zx_1 \oplus vx_2 \oplus zx_2 \oplus x_2 \oplus v \oplus z \quad (4.22)$$

$$y_1 \oplus y_2 = vx_2 \oplus zx_2 \oplus x_1 \quad (4.23)$$

We see that the first subset of the elements $\mathbf{F}_{2/2}$ is preferable as primitives for constructing RCOs due to its higher nonlinearity (NL) value. In the case of the switchable CSPNs containing elements $\mathbf{F}_{2/2}$, the notion of symmetric distribution of the controlling bits relates to the following definition:

Definition 4.5′. *Distribution of the controlling bits of the* $(V, Z) = (V_1, Z_1, V_2, Z_2, \ldots, V_{s/2}, Z_{s/2})$ *vector is called symmetric if for all* $j = 1, \ldots, [s/2]$ *the following relations hold:* $V_j = V_{s-j+1}$ *and* $Z_j = Z_{s-j+1}$.

4.3 Block Ciphers with Switchable DDOs

4.3.1 Updating the DDP-Based Block Ciphers

For efficient firmware and software implementation, the iterated DDP-based ciphers Cobra-F64a, Cobra-F64b, and Cobra-S128 have been proposed [48], in which the variable permutations are performed with the switchable DDP box $\mathbf{P}^{(V,e)}_{32/96}$ having the same construction structure as the RCO box $\mathbf{R}^{(V,e)}_{32/96}$, except that the switching elements $\mathbf{P}_{2/1}$ are used instead of the CEs $\mathbf{R}_{2/1}$. The mentioned ciphers are also efficient for hardware implementation. Moreover, the hardware implementation efficacy can be significantly increased using one of the SDDO boxes $\mathbf{U}^{(V,e)}_{32/96}$, $\mathbf{Z}^{(V,e)}_{32/96}$, $\mathbf{R}^{(V,e)}_{32/96}$, $\mathbf{Q}^{(V,e)}_{32/96}$, or $\mathbf{\Phi}^{(V,e)}_{32/96}$ instead of the box $\mathbf{P}^{(V,e)}_{32/96}$ and the XOR operations instead of the modulo 2^{32} addition and subtraction operations. Such updating of the known ciphers is similar to that used in Section 3.5.3 in Chapter 3 and provides for reduction of the number of encryption rounds. The round transformations of the updated ciphers are shown in Figure 4.10. All ciphers use the 128-bit secret key $K = (K_1, K_2, K_3, K_4)$, where $\forall\, i\; K_i \in \{0, 1\}^{32}$. The key scheduling is very simple, i.e., 32-bit subkeys K_i are directly used during data ciphering. In each round, two subkeys are used as one of the 64-bit round keys $(Q^{(1,e)}_j, Q^{(2,e)}_j)$, where $j = 1, \ldots, R + 1$,

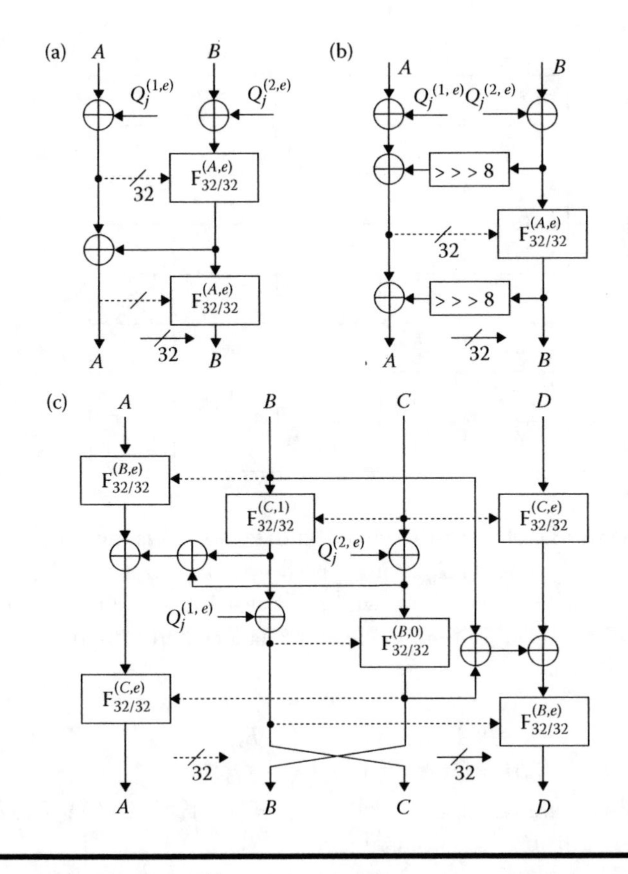

Figure 4.10 Procedure Crypt$^{(e)}$ in the updated versions of Cobra-F64a (*a*), Cobra-F64b (*b*), and Cobra-S128 (*c*).

R denotes the number of the transformation rounds, and $e = 0$ ($e = 1$) denotes the encryption (decryption) process. Correspondence between the secret key and the round subkeys is defined by Table 4.2 and the following formulas for Cobra-F64a and Cobra-F64b:

$$(\mathcal{Q}_1^{(1,1)}, \mathcal{Q}_{R+1}^{(1,1)}) = (\mathcal{Q}_{R+1}^{(1,0)}, \mathcal{Q}_1^{(1,0)}), (\mathcal{Q}_1^{(2,1)}, \mathcal{Q}_{R+1}^{(2,1)}) = (\mathcal{Q}_{R+1}^{(2,0)}, \mathcal{Q}_1^{(2,0)}) \tag{4.24}$$

$$(\mathcal{Q}_j^{(1,1)}, \mathcal{Q}_j^{(2,1)}) = (\mathcal{Q}_{R-j+2}^{(2,0)}, \mathcal{Q}_{R-j+2}^{(1,0)}), \forall j = 2,\ldots,R \tag{4.25}$$

For Cobra-S128 the following correspondence is specified:

$$(\mathcal{Q}_j^{(1,1)}, \mathcal{Q}_j^{(2,1)}) = (\mathcal{Q}_{R-j+1}^{(2,0)}, \mathcal{Q}_{R-j+1}^{(1,0)}), \forall j = 1,\ldots,R \tag{4.26}$$

Table 4.2 Key scheduling in Cobra-F64a, Cobra-F64b, and Cobra-S128

j	$Q_j^{(1,0)}$	$Q_j^{(2,0)}$	j	$Q_j^{(1,0)}$	$Q_j^{(2,0)}$	j	$Q_j^{(1,0)}$	$Q_j^{(2,0)}$
1	K_1	K_4	8	K_3	K_4	15	K_2	K_3
2	K_2	K_3	9	K_1	K_2	16	K_3	K_4
3	K_3	K_1	10	K_2	K_3	17	K_1	K_2
4	K_4	K_2	11	K_4	K_1	18	K_4	K_1
5	K_2	K_3	12	K_3	K_2	19	K_3	K_4
6	K_1	K_2	13	K_1	K_3	20	K_1	K_2
7	K_4	K_1	14	K_4	K_1	—	—	—

In Cobra-F64a and Cobra-F64b, the 64-bit data block X is divided into two 32-bit subblocks A and B. The data ciphering is performed in two stages: (1) R rounds with e-dependent procedure $\mathbf{Crypt}^{(e)}$ shown in Figure 4.10 and (2) final transformation. For the updated versions of both ciphers, the data ciphering algorithm can be represented as follows:

For $j = 1$ to $R - 1$, do: $\{(A, B) := \mathbf{Crypt}^{(e)}(A, B, Q_j^{(1,e)}, Q_j^{(2,e)}); (A, B) := (B, A)\}$.
For $j = R$, do: $\{(A, B) := \mathbf{Crypt}^{(e)}(A, B, Q_j^{(1,e)}, Q_j^{(2,e)})\}$.
Perform final transformation as follows: $\{Y = (Y_l, Y_h) := (A \oplus Q_{R+1}^{(1,e)}, B \oplus Q_{R+1}^{(2,e)})\}$, where Y is the 64-bit output data block.

In Cobra-S128, the 128-bit input block X is divided into four 32-bit subblocks A, B, C, and D, and data ciphering is performed using procedure $\mathbf{Crypt}^{(e)}$ (Figure 4.10c) as follows:

1. Perform initial transformation:

$$\{(A,B,C,D) := (A \oplus Q_1^{(1,e)}, B \oplus Q_2^{(1,e)}, C \oplus Q_3^{(1,e)}, D \oplus Q_4^{(1,e)})\}.$$

2. For $j = 1$ to $R-1$ do:

$$\{(A,B,C,D) := \mathbf{Crypt}^{(e)}(A,B,C,D,Q_j^{(1,e)},Q_j^{(2,e)}); (A,B,C,D) := (B,A,D,C)\}.$$

3. Then, do as follows:

$$\{(A,B,C,D) := \mathbf{Crypt}^{(e)}(A,B,C,D, Q_R^{(1,e)}, Q_R^{(2,e)})\}.$$

4. Perform final transformation:

Table 4.3 Updating some DDP-based ciphers

Cipher	Data block size, bits	$t_{\oplus}$[a], arb.un.	SDDO boxes		Secure number of rounds	
			Initial	New[b]	Initial	New
Cobra-F64a	64	14	$\mathbf{P}^{(V,e)}_{32/96}$	$\mathbf{R}^{(V,e)}_{32/96}$	16	12
Cobra-F64a	64	14	$\mathbf{P}^{(V,e)}_{32/96}$	$\mathbf{Q}^{(V,e)}_{32/96}$	16	10
Cobra-F64b	64	9	$\mathbf{P}^{(V,e)}_{32/96}$	$\mathbf{Q}^{(V,e)}_{32/96}$	20	14
Cobra-F64b	64	9	$\mathbf{P}^{(V,e)}_{32/96}$	$\mathbf{\Phi}^{(V,e)}_{32/96}$	20	12
Cobra-S128	128	21	$\mathbf{P}^{(V,e)}_{32/96}$	$\mathbf{Q}^{(V,e)}_{32/96}$	12	10
Cobra-S128	128	21	$\mathbf{P}^{(V,e)}_{32/96}$	$\mathbf{\Phi}^{(V,e)}_{32/96}$	12	8

[a] Time delay of the round transformation.
[b] The **R**-, **Q**-, and **Φ**-boxes are constructed using the (i,e), (i,j), and (h,f,e,j) CEs, correspondingly.

$$\{Y = (A,B,C,D) := (A \oplus Q_{\mathrm{R}}^{(2,e)},\ B \oplus Q_{\mathrm{R}-1}^{(2,e)},\ C \oplus Q_{\mathrm{R}-2}^{(2,e)},\ D \oplus Q_{\mathrm{R}-3}^{(2,e)})\}.$$

Some variants of the updated ciphers are presented in Table 4.3. The security estimation results against differential attack are presented in Table 4.4, where $P(N)$ denotes the probability with which the iterative differences pass through N rounds. We see that replacement of the DDP operations by DDO ones significantly reduces the value $P(2)$, and a minimum number of the encryption rounds is required to make the encryption procedure indistinguishable from the random transformation. Security comparison of the initial and updated variants of ciphers is shown in Table 4.4.

4.3.2 Hawk-64: A Cipher Based on Switchable Data-Driven Operations

Controllability property of the operations based on permutation networks (PNs) and CSPNs can be efficiently used in the block cipher design oriented to cheap hardware implementation if the transformed data are used to specify the current values of the controlling vector. To minimize the implementation cost we can use simple key scheduling; however, in the iterated ciphers that perform encryption and decryption with the same algorithm, there are some problems connected with

Table 4.4 Security comparison against DCA

Cipher	N (# rounds)	SDDO boxes	Difference of the z-round DC	$P(z)$	$P(N)$
Cobra-F64a	16	**P**-type [48]	(Δ_1^A, Δ_0^B)	$P(3) = 2^{-21}$	$< 2^{-100}$
	12	**R**-type	(Δ_1^A, Δ_0^B)	$P(1) < 2^{-8}$	$< 2^{-96}$
	10	**Q**-type	(Δ_1^A, Δ_0^B)	$P(2) < 2^{-20}$	$< 2^{-100}$
Cobra-F64b	20	**P**-type [48]	(Δ_0^A, Δ_1^B)	$P(2) = 2^{-12}$	$\approx 2^{-120}$
	14	**Q**-type	(Δ_1^A, Δ_0^B)	$P(2) < 2^{-18}$	$< 2^{-126}$
	12	**Φ**-type	(Δ_1^A, Δ_0^B)	$P(2) < 2^{-24}$	$< 2^{-144}$
Cobra-S128	12	**P**-type [48]	$(\Delta_1^A,\Delta_0^B,\Delta_0^C,\Delta_1^D)$	$P(2) = 2^{-32}$	$< 2^{-190}$
	10	**R**-type	$(\Delta_1^A,\Delta_0^B,\Delta_0^C,\Delta_1^D)$	$P(2) < 2^{-50}$	$< 2^{-200}$
	8	**Φ**-type	$(\Delta_1^A,\Delta_0^B,\Delta_0^C,\Delta_1^D)$	$P(2) < 2^{-56}$	$< 2^{-224}$

direct use of nontransformed parts of the secret key. One of the problems relates to the weak keys that define the sameness of the encryption and decryption procedures. For example, if 32-bit subkeys K_i ($i = 1, \ldots, 4$) of the 128-bit secret key $K = (K_1, K_2, K_3, K_4)$ are used directly in the round transformation and the encryption and decryption procedures differ only in key scheduling, then, for arbitrary X, the key $K = (X, X, X, X)$ is weak; i.e., the transformation procedure is involution, because transpositions of equal subkeys do not introduce any change in the transformation process. Usually, the portion of the weak keys is extremely small; however, they put limitation on using such ciphers to compute keyless data integrity checksums (for example, MAC).

To avoid this problem, we can use ciphers in which enciphering and deciphering algorithms are different, or use the secret-key processing on the fly, but such an approach significantly increases the hardware implementation cost. We can use preprocessing the key, but this essentially decreases the performance when the keys are often changed. It appears that using of SDDOs represents the appropriate solution to the problem. Indeed, if $e = 0$ is set, then we have some direct transformation that is not an involution for any possible value of the secret key. If $e = 1$ is set, then we have respective inverse transformation. Actually, we have two different algorithms for encryption and decryption such as in the former case. However, in the latter case, the hardware implementation is sufficiently more efficient.

The cipher Hawk-64 represents a variant of designs based on SDDOs. It operates on 64-bit data blocks and uses 128-bit key $K = (K_1, K_2, K_3, K_4)$, represented as the concatenation of four 32-bit keys. Figure 4.11 presents the design of the eight-round cipher Hawk-64. A feature of this cipher is the using of the same key

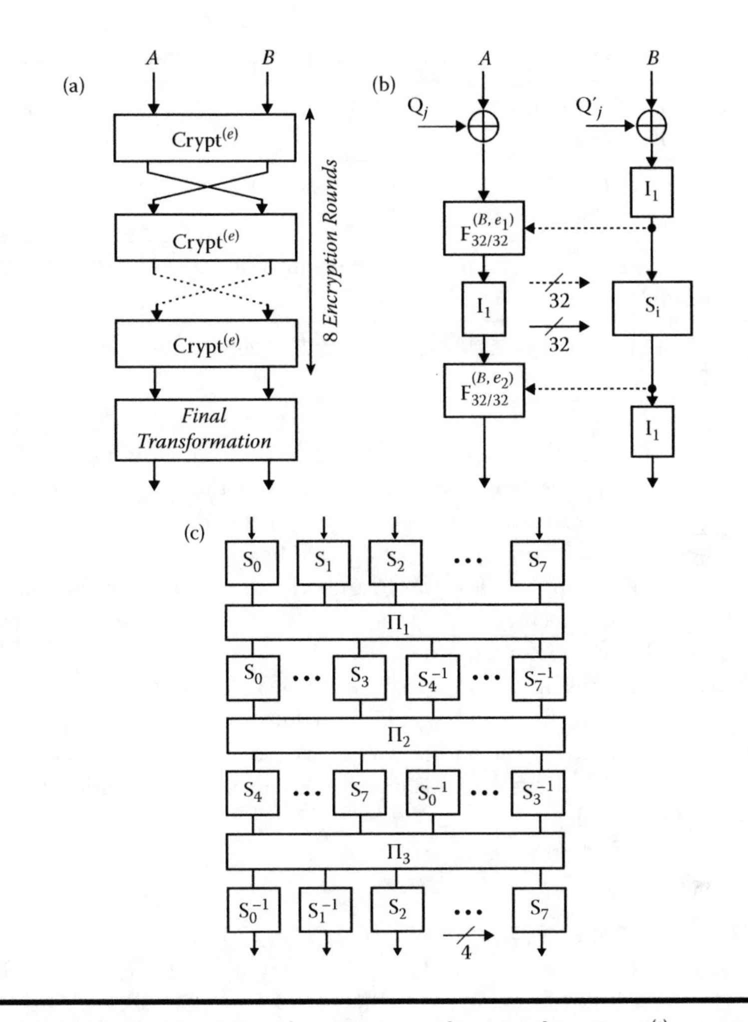

Figure 4.11 Hawk-64: (*a*) iterative structure; (*b*) procedure Crypt$^{(e)}$; (*c*) topology of the S_i operation.

scheduling for encryption and decryption. This is due to the symmetry of the key scheduling. Of course, we can use nonsymmetric key scheduling for encryption and respective inverse scheduling for decryption, but the used case reduces the implementation cost, especially in the case of the iterated implementation architectures. Another feature is the using of the scheduling of the switching bits e_1 and e_2 (their values depend on the round number). This prevents the periodicity of the transformation procedure in case of selecting a key such as $K = (X, X, X, X)$.

In the round transformation of Hawk-64, the right branch represents some CSPN performing two SDDOs $F_{32/32}^{(B,e_1)}$ and $F_{32/32}^{(B',e_2)}$ where the right data subblock B is used as the controlling vector. The operations $F_{32/32}^{(B,e_1)}$ and S_i are performed in parallel, therefore the vector B' is formed before the beginning of the operation $F_{32/32}^{(B',e_2)}$

Table 4.5 Fixed permutations of the round transformation of Hawk-64

$\mathbf{I}_1$	(1)(2,9)(3,17)(4,25)(5)(6,13)(7,21)(8,29)(10)(11,18)(12,26)(14)(15,22)(16,30)(19)(20,27)(23)(24,31)(28)(32)
Π_1	(1)(2,5)(3,17)(4,21)(6)(7,18)(8,25)(9)(10,13)(11,22)(12,29)(14)(15,26)(16,30)(19)(20,23)(24)(27)(28,31)(32)
Π_2	(1)(2,5)(3,9)(4,13)(6)(7,10)(8,14)(11)(12,15)(16)(17)(18,21)(19,25)(20,29)(22)(23,26)(24,30)(27)(28,31)(32)
Π_3	(1,3,19,17)(2,7,20,21)(4,23,18,5)(6,11,27,22)(8,24,25,9)(10,15,28,29)(12,31,26,13)(14,16,32,30)

(for example, when the field programmable gate array (FPGA) implementation of the latency of the S_i operation is smaller than that of the $F_{32/32}^{(B,e_1)}$ operation). The right branch represents the switchable permutation network (SPN) containing four layers of the four-bit S boxes. The used $S_0, \ldots, S_7$ boxes and their inverses $S_0^{-1}, \ldots, S_7^{-1}$ are described in Chapter 3, Tables 3.16 and 3.17, respectively. The permutations $\mathbf{I}_1$, Π_1, Π_2, and Π_3 are specified in Table 4.5. The SPN performs the transformation that is an involution; therefore, it is denoted as S_i.

In general, the transformation procedure of Hawk-64 is described as $Y = \mathbf{T}^{(e)}(X, K)$, where $X = (A, B)$ is the input data block (plaintext while encrypting or ciphertext while decrypting), Y is the output data block, $X, Y \in \{0, 1\}^{64}$, $\mathbf{T}^{(e)}$ is the transformation function, and $e \in \{0, 1\}$ is a parameter defining encryption ($e = 0$) or decryption ($e = 1$) mode.

The data transformation procedure of Hawk-64 comprises the following generalized steps:

1. For $j = 1$ to 7, do: $\{(L, R) \leftarrow \mathbf{Crypt}^{(e)}(L, R, Q_j, Q'_j); (L, R) \leftarrow (R, L)\}$.
2. Transform the (L, R) data block: $(L, R) \leftarrow \mathbf{Crypt}^{(e)}(L, R, Q_8, Q'_8)$.
3. Perform final transformation: $\{(L, R) \leftarrow (L \oplus Q_9, R \oplus Q'_9)\}$.

The final transformation (FT) is implemented as simple XORing data subblocks with respective subkeys. It is used to define the same algorithm for both the encryption and the decryption procedures. The $F_{32/32}^{(B,e_1)}$ and $F_{32/32}^{(B',e_2)}$ boxes are constructed using the (i,j) CEs. Topology of these operations is described in Section 4.2. The switching bits e_1 and e_2 depend on the bit e and round number as follows:

$$e_1 = e' \oplus e \text{ and } e_2 = e'' \oplus e \tag{4.27}$$

where e' and e'' are specified as in Table 4.6. The permutational involution $\mathbf{I}_1$ in the left and right branches of the round transformation is the same as the one used to connect the cascade of the $\mathbf{F}_{8/12}$ boxes with the cascade of the $\mathbf{F}_{8/12}^{-1}$ boxes in the box $\mathbf{F}_{32/96}$.

Table 4.6 Key scheduling and specification of the switching bits e′ and e″

$j =$	1	2	3	4	5	6	7	8	9[a]
$Q_j =$	K_1	K_2	K_3	K_4	K_4	K_2	K_2	K_3	K_1
$Q'_j =$	K_3	K_3	K_2	K_2	K_4	K_4	K_3	K_2	K_3
e′ =	1	0	1	1	0	1	0	0	—
e″ =	0	0	1	0	1	1	0	1	—

[a] $j = 9$ corresponds to final transformation.

Considering the possible formation mechanisms of the iterative differential characteristic (DC) with difference (Δ_1^A, Δ_0^B), we can estimate the probability that this difference passes through two rounds of Hawk-64 as $P(2) < 2^{-24} \cdot P_S$, where $P_S < 2^{-5}$ is the probability that difference Δ_1^B passes through the S_i operation. Five transformation rounds with Hawk-64 are sufficient to thwart linear cryptanalysis (LCA) and differential cryptanalysis (DCA). Three additional rounds are added to obtain a sufficiently large value for the security margin. Indeed, for Hawk-64 we have $100\% \cdot (R - R_{min})/R_{min} = 60\%$, where R and R_{min} are the specified and the minimum secure number of rounds, respectively.

4.4 Designs of the Bit Permutation Instruction for General Purpose Processors

Different algorithms widely used in practice include the bit permutation operations. We can attribute such algorithms to the following two classes: (1) noncryptographic and (2) cryptographic. Usually, in the algorithms of the first type, fixed permutations are used, and different algorithms use their variants. Therefore, in case of noncryptographic algorithms, we should consider the possibility of implementing arbitrary fixed permutations. Both the fixed and the variable permutations are used in cryptographic algorithms. In this case, we deal with the new type of permutation operations that are used as main primitive in the DDP-based block ciphers considered in detail in Chapter 2. Algorithms using bit permutation operations (fixed or variable) have comparatively low performance in case of software implementation, except the ones using fixed or variable rotation of the bit strings.

To increase the performance of the considered algorithms, it is important to develop suitable approaches for embedding a universal bit permutation instruction (BPI) in the general purpose processors. Under universal BPI, we understand a BPI that allows to perform both the variable (data-driven) permutations and the arbitrary fixed permutations. To develop a circuit implementing both types of the bit permutations, we need to solve the following two problems:

- Implementation of arbitrary fixed permutations requires us to use a sufficiently large controlling vector; for example, to permute a 32-bit (64-bit) string with the PN of the maximum order, the controlling vector should have the size of 144 bits (352 bits), i.e., we have to decide where to store the controlling values.
- Implementation of the data-driven permutations requires us to change the controlling value rapidly (as rapidly as possible, because this defines the encryption performance). We can gain the maximum rapidity if the registers containing the data subblocks are used as controlling registers. During software implementation, we usually have 32- or 64-bit controlling data subblocks (depending on the data bus size of the processor); therefore, only a very small part of possible variants of permutations can be implemented. Fast implementation of the variable permutations contradicts the implementation of the arbitrary prescribed permutations. If some controlling register having a large size is used as an internal part of the circuit implementing BPI, then each data-driven permutation will be performed at least in six to seven cycles because the controlling value should be loaded into the controlling register. Thus, we should decide how to perform quickly both the variable and the arbitrary fixed permutations.

In the microcontrollers oriented to special applications, some specialized architectures of the BPI can be applied. For cryptographic applications, we can use the PN with symmetric topology and implement the switchable PBIs, for example, $\mathrm{P}^{(V,e)}_{32/96}$ ($h = 2$), $\mathrm{P}^{(V,e)}_{32/128}$ ($h = 8$), or $\mathrm{P}^{(V,e)}_{32/144}$ ($h = 32$). In these examples, we have a sufficiently large controlling value V, but it is an internal value that is formed with the internal extension box that is implemented as a simple wiring. Therefore, the formation of the controlling value V, depending on the data subblock value contained in the external controlling register, does not introduce any time delay. In this case, only a small part of all possible variants of the bit permutations is implemented; however, the data-driven permutation can be implemented in one cycle. Moreover, in different cases more efficient cryptographic operations, such as $\mathrm{R}^{(V,e)}_{32/96}$, $\mathrm{Q}^{(V,e)}_{32/128}$, or $\Phi^{(V,e)}_{32/288}$ can be reasonably imbedded in cryptographic microcontrollers.

For noncryptographic applications, we can use the PN having the maximum order as the core unit of the BPI implementing arbitrary fixed permutations in one cycle. The required long controlling vector should be stored in some internal controlling register that is to be preloaded with the corresponding controlling value. Such initialization process requires five or six cycles for 32- or 64-bit microcontrollers, respectively. After the controlling value is set, the corresponding fixed bit permutation is performed in one cycle.

In case of general purpose processors, we should take into account the algorithms corresponding to both the types of bit permutations. Indeed, in this case the universal architecture of the BPI is most attractive.

The development of BPIs for both the general purpose processor and the specialized microcontrollers is of practical interest. In the literature [24–26,48], several design variants of the BPI for embedding in the general purpose processor have been proposed. However, the proposed designs can be considered only as specialized rather than universal. In the following subsections, we consider the specialized and universal architectures of the BPI.

4.4.1 Design of the BPI for Cryptographic Applications

From the point of view of construction of encryption algorithms and hash functions, it is not necessary to perform some prescribed bit permutations. It is sufficient to specify the permutations that provide good avalanche effect and sufficient uniformity of the probability distribution of the bit transpositions. The $P^{(V,e)}_{32/96}$ operation used in the design of COBRA-F64a and COBRA-F64b block ciphers [48] is an example of such command. An alternative variant of the specialized BPI is the $P^{(L,e)}_{32/128}$ operation, the design of which is described hereunder. The $P^{(L,e)}_{32/128}$ operation is analogous to $P^{(V,e)}_{32/96}$ operation. The difference between these two operations corresponds to more number of active layers used in $P^{(L,e)}_{32/128}$ and to different distributions of the controlling data subblock bits.

The core element of the $P^{(L,e)}_{32/128}$ operation design is the $\mathbf{P}_{32/128}$ box having symmetric topology and the order $h = 8$. It can be represented as superposition

$$\mathbf{P}_{32/128} = \mathbf{P}^{(U)}_{32/64} \circ \pi_{\text{inv}} \circ (\mathbf{P}^{-1}_{32/64})^{(U')} \tag{4.28}$$

where $\mathbf{P}_{32/64} = \mathbf{L}^{(V1)} \circ \pi_{(5)} \circ \mathbf{L}^{(V2)} \circ (\pi_{(4)} || \pi_{(4)}) \circ \mathbf{L}^{(V3)} \circ (\pi_{(3)} || \pi_{(3)} || \pi_{(3)} || \pi_{(3)}) \circ \mathbf{L}^{(V4)}$ and $\mathbf{P}^{-1}_{32/64} = \mathbf{L}^{(V5)} \circ (\pi^{-1}_{(3)} || \pi^{-1}_{(3)} || \pi^{-1}_{(3)} || \pi^{-1}_{(3)}) \circ \mathbf{L}^{(V6)} \circ (\pi^{-1}_{(4)} || \pi^{-1}_{(4)}) \circ \mathbf{L}^{(V7)} \circ \pi^{-1}_{(5)} \circ \mathbf{L}^{(V8)}$. The permutational involution π_{inv} represents the concatenation of eight linking permutations $\pi_{(2)}$, i.e., $\pi_{\text{inv}} = \pi_{(2)} || \pi_{(2)} || \pi_{(2)} || \pi_{(2)} || \pi_{(2)} || \pi_{(2)} || \pi_{(2)} || \pi_{(2)}$. The direct description of the π_{inv} fixed permutation is as follows:

(1)(2,3)(4)(5)(6,7)(8)(9)(10,11)(12)(13)(14,15)(16)(17)(18,19)
(20)(21)(22,23)(24)(25)(26,27)(28)(29)(30,31)(32).

Design of the cryptography-oriented BPI $P^{(L,e)}_{32/128}$ is shown in Figure 4.12. Two halves of the controlling data subblock $L = (L_l, L_h)$ are fed to the input of the transposition box $\mathbf{P}^{(e)}_{32/1}$. The input of the $\mathbf{P}^{(e)}_{32/1}$ box is divided into 16-bit left and 16-bit right inputs. The transposition box $\mathbf{P}^{(e)}_{32/16}$ contains 16 parallel boxes $\mathbf{P}^{(e)}_{2/1}$ controlled with the same bit, e. The left (right) inputs of the $\mathbf{P}^{(e)}_{2/1}$ boxes correspond to the left (right) 16-bit input of the $\mathbf{P}^{(e)}_{32/16}$ box. If the input vector of the $\mathbf{P}^{(e)}_{32/1}$ box is $L = (L_l, L_h)$, then, at the output of $\mathbf{P}^{(e)}_{32/1}$, we have $(U, U') = (L_l, L_h)$ if $e = 0$, or $(U, U') = (L_h, L_l)$ if $e = 1$. Two 16-bit output vectors U and U' of the $\mathbf{P}^{(e)}_{32/1}$ box are fed to the extension boxes $\mathbf{E}$ and $\mathbf{E}'$, correspondingly. The boxes $\mathbf{E}$ and $\mathbf{E}'$ produce the controlling vectors (V_1, V_2, V_3, V_4) and (V_5, V_6, V_7, V_8), correspondingly. The extension boxes define the following symmetric distribution of the controlling bits:

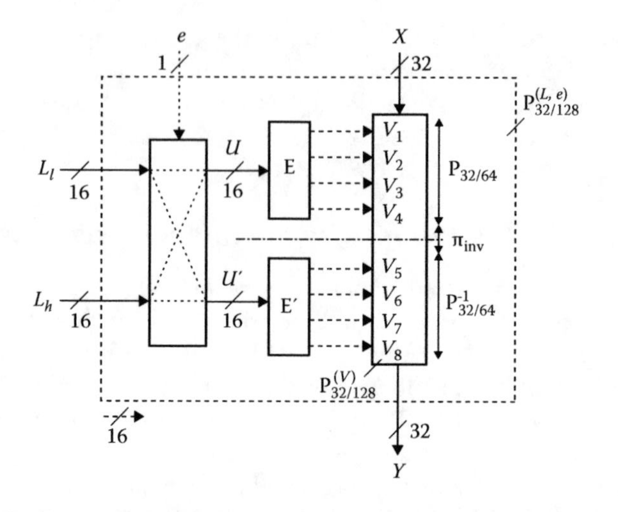

Figure 4.12 Design of the bit permutation instruction $\mathbf{P}^{(L,e)}_{32/128}$ suitable for embedding in microcontrollers.

$$V_1 = (u_1, u_2, u_3, u_4, u_1, u_2, u_3, u_4, u_1, u_2, u_3, u_4, u_1, u_2, u_3, u_4);$$
$$V_2 = (u_8, u_7, u_6, u_5, u_5, u_6, u_7, u_8, u_8, u_7, u_6, u_5, u_5, u_6, u_7, u_8);$$
$$V_3 = (u_9, u_{10}, u_{11}, u_{12}, u_9, u_{10}, u_{11}, u_{12}, u_9, u_{10}, u_{11}, u_{12}, u_9, u_{10}, u_{11}, u_{12});$$
$$V_4 = (u_{13}, u_{14}, u_{15}, u_{16}, u_{16}, u_{15}, u_{14}, u_{13}, u_{13}, u_{14}, u_{15}, u_{16}, u_{16}, u_{15}, u_{14}, u_{13});$$
$$V_5 = (u'_{13}, u'_{14}, u'_{15}, u'_{16}, u'_{16}, u'_{15}, u'_{14}, u'_{13}, u'_{13}, u'_{14}, u'_{15}, u'_{16}, u'_{16}, u'_{15}, u'_{14}, u'_{13});$$
$$V_6 = (u'_9, u'_{10}, u'_{11}, u'_{12}, u'_9, u'_{10}, u'_{11}, u'_{12}, u'_9, u'_{10}, u'_{11}, u'_{12}, u'_9, u'_{10}, u'_{11}, u'_{12});$$
$$V_7 = (u'_8, u'_7, u'_6, u'_5, u'_5, u'_6, u'_7, u'_8, u'_8, u'_7, u'_6, u'_5, u'_5, u'_6, u'_7, u'_8);$$
$$V_8 = (u'_1, u'_2, u'_3, u'_4, u'_1, u'_2, u'_3, u'_4, u'_1, u'_2, u'_3, u'_4, u'_1, u'_2, u'_3, u'_4).$$

For such distribution of the controlling bits, each bit l_i of the controlling data subblock defines four bits of the vector $V = (V_1, V_2, \ldots, V_8)$, and for all values L, each bit of the permuted data subblock X is moved depending on eight different bits of L. If $e = 0$, then $U = L_l$ and $U' = L_h$; if $e = 1$, then $U = L_h$ and $U' = L_l$. Therefore, due to the symmetric distribution of the bits x_i and x'_i for $i = 1, 2, \ldots, 16$, we have $\mathrm{P}^{(L,e)}_{32/128} = \left(\mathrm{P}^{(L,e\oplus 1)}_{32/128}\right)^{-1}$ for arbitrary given value $L = (L_l, L_h)$, i.e., we have switchable BPI performing variable bit permutations.

4.4.2 Design of the BPI for Non-Cryptographic Applications

The BPI oriented to noncryptographic applications should perform different variants of specified fixed permutations; therefore, it should be designed on the basis of the maximum order PN, for example, on the basis of the $\mathbf{P}_{32/144}$ boxes in case of

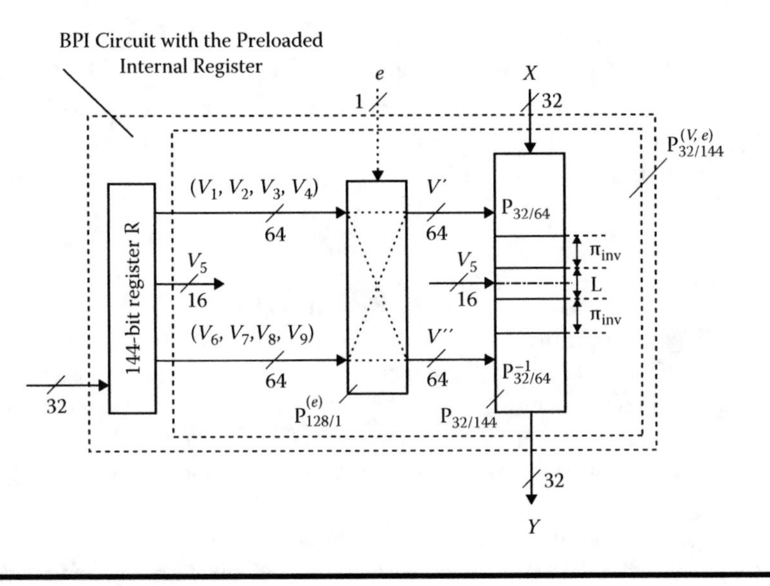

Figure 4.13 Design of the BPI oriented to fast implementation of arbitrary fixed permutations.

32-bit microcontrollers. The $\mathbf{P}_{32/144}$ box has symmetric structure and can be represented as the following involution (see Figure 4.13):

$$\mathbf{P}_{32/144} = \mathbf{P}_{32/64}^{(V')} \circ \pi_{\text{inv}} \circ \mathbf{L}^{(V5)} \circ \pi_{\text{inv}} \circ (\mathbf{P}_{32/64}^{-1})^{(V'')} \tag{4.29}$$

where V_5, $V' = (V_1', V_2', V_3', V_4')$ and $V'' = (V_1'', V_2'', V_3'', V_4'')$ are independent vectors. In some cases, in the frame of the same algorithm, two mutually inverse permutations are performed frequently; therefore, it is reasonable to implement the switchable BPI architecture, which is presented in Figure 4.13. Using such a design we can implement all possible permutations. The internal preloaded register serves to store the 144-bit controlling value $V = (V_1, V_2, V_3, V_4, V_5, V_6, V_7, V_8, V_9)$ that specifies the required fixed permutation. (For any given bit permutation, one can easily calculate the required value V.) After the register is preloaded, the needed permutation of the input bit string X is performed quickly in one cycle. If necessary, two mutually inverse permutations can be performed in two cycles.

In a study by Waksman [68], a topology of the maximum order PN is proposed that uses less number of switching elements $\mathbf{P}_{2/1}$ (129 against 144) and shorter controlling vector than the $\mathbf{P}_{32/144}$ box; however, the PN topology is not symmetric and is not suitable for efficient implementation of the switchable BPI.

Using symmetric $\mathbf{P}_{32/144}$ boxes and transposition box $\mathbf{P}_{128/1}^{(e)}$, we get switchable BPI. The transposition box $\mathbf{P}_{128/1}^{(e)}$ contains 64 parallel $\mathbf{P}_{2/1}^{(e)}$ boxes controlled with the switching bit e. The left (right) inputs of the $\mathbf{P}_{2/1}^{(e)}$ boxes correspond to the left (right) 64-bit input of the $\mathbf{P}_{128/1}^{(e)}$ box. If the input vector of the $\mathbf{P}_{128/1}^{(e)}$ box is $(V_1, V_2, V_3, V_4,$

V_6, V_7, V_8, V_9), then, at the output of $\mathbf{P}^{(e)}_{128/1}$ we have (V', V'') where $V' = (V_1, V_2, V_3, V_4)$ and $V'' = (V_6, V_7, V_8, V_9)$ if $e = 0$, or $V' = (V_9, V_8, V_7, V_6)$ and $V'' = (V_4, V_3, V_2, V_1)$ if $e = 1$. Two 16-bit output vectors V' and V'' of the $\mathbf{P}^{(e)}_{128/1}$ box are fed to the $\mathbf{P}_{32/64}$ and $\mathbf{P}^{-1}_{32/64}$ boxes, correspondingly. It is easy to see that for arbitrary given value loaded in the register **R**, we have

$$\mathrm{P}^{(V,\,e)}_{32/144} = \left(\mathrm{P}^{(V,\,e\oplus 1)}_{32/144}\right)^{-1} \tag{4.31}$$

The number of all possible permutations is equal to $n!$, where n is the size of the data bus ($n = 32$ or $n = 64$); therefore, it is practically impossible to calculate and store the table specifying the controlling vector values for all permutations. However, the controlling vector values defining the most frequently used permutations can be precalculated and written in the respective tables. To specify the arbitrary new permutation, one can elaborate the program of calculation of the corresponding controlling value.

Naturally, the considered BPI increases the software performance of the DDP-based ciphers if the data-dependent permutations (DDPs) are implemented with the BPI. However, the published variants of the DDP-based ciphers use the CP boxes that implement other types of DDP operations than the BPI based on maximum order PNs. This means that performance increase can be gained for the modified version of the known DDP-based encryption algorithms in which the DDP operations are performed with the considered BPI. The modified variants of the ciphers will be significantly faster in software than the initial versions. Besides, the CP boxes of maximum order are more efficient as cryptographic primitives. However, the controlling vector in the DDP-based ciphers repeatedly changes during the encryption process, making it necessary to reload the register **R** in accordance with the transformation of the controlling data subblock value. This brings the necessity of several machine cycles for performing each DDP operation resulting in significant reduction of the cryptographic transformation speed against performance in hardware. Instead of complete reloading of the **R** register, we can define the updating of only 32 bits in the register. In this case only two cycles are needed to perform one DDP operation; however, it leads to a significant deterioration of the DDP in cryptographic operation.

In the next subsection, we consider the architecture of the universal BPI oriented to fast implementation of both the DDP and the arbitrary fixed permutations.

In general, the switching $\mathrm{P}^{(e)}_{128/1}$ box can be excluded from the design presented in Figure 4.13 because the controlling vector that defines the inverse permutation can be calculated simultaneously with the vector defining the direct permutation. However, this leads to delay equal to several machine cycles because in such a case, reversing the bit permutations operation requires reloading the register **R**. For algorithms including the periodic usage of mutually inverse permutations, this delay will reduce the performance. Therefore, the design that provides for the possibility of fast reversal of permutation operations is more attractive.

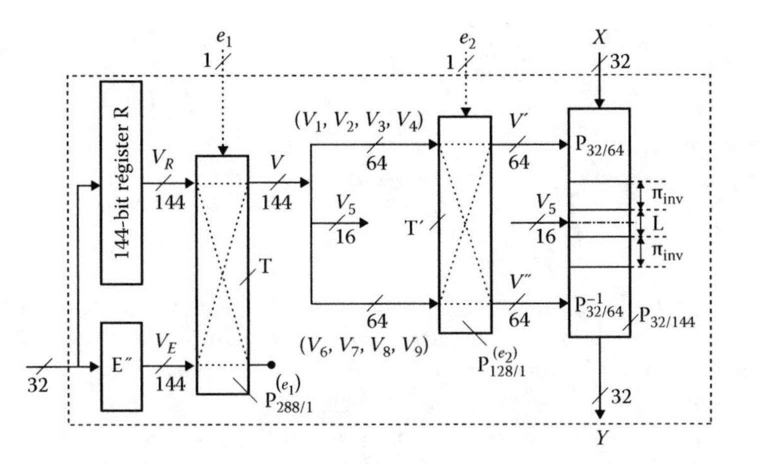

Figure 4.14 Design of the universal BPI $P^{(V, e_1, e_2)}_{32/144}$.

4.4.3 Architecture of the Universal BPI

In the design of the universal BPI, we should provide for the following possibilities:

- Fast performance the arbitrary fixed permutations
- Fast reversal of the current fixed permutation of the arbitrary type
- Fast performance of the data-dependent permutations
- Fast reversal of the data-dependent permutations

These problems can be solved by combining two designs presented in Figures 4.12 and 4.13. To unite the functions of both designs, we can use the additional transposition box $P^{(e_1)}_{288/1}$ (box **T**) as is shown in Figure 4.14. The box $P^{(e_1)}_{288/1}$ connects the output of the register **R** with the input of the switching box $P^{(e_2)}_{128/1}$ (box **T′**) if $e_1 = 0$, or connects the output of the extension box **E″** with the input of the switching box $P^{(e_2)}_{128/1}$ if $e_1 = 1$. Thus, in the former case, we can quickly perform the arbitrary fixed permutations and reverse them by inverting the bit e_2. In the latter case, we can quickly perform data-dependent permutations (DDPs) and reverse them also by inverting the bit e_2. The box **E″** differs from the boxes **E** and **E′** used in Figure 4.12. Actually, the size of the output (input) of the first extension box is equal to 144 bits (32 bits), whereas the **E** and **E′** boxes have the 64-bit output and 16-bit input.

The box **E″** can be defined with the following relations that satisfy criteria 2.1 and 2.2 mentioned in Chapter 2, Section 2.3:

$$V_1 = (l_1, l_2, l_1, l_2, l_3, l_4, l_3, l_4, l_3, l_4, l_3, l_4, l_1, l_2, l_1, l_2)$$
$$V_2 = (l_5, l_6, l_6, l_5, l_7, l_7, l_8, l_8, l_8, l_8, l_7, l_7, l_5, l_5, l_6, l_6)$$

$$V_3 = (l_9, l_{10}, l_{13}, l_{14}, l_9, l_{10}, l_{13}, l_{14}, l_{10}, l_9, l_{13}, l_{14}, l_{10}, l_9, l_{13}, l_{14})$$
$$V_4 = (l_{11}, l_{12}, l_{11}, l_{12}, l_{15}, l_{16}, l_{15}, l_{16}, l_{15}, l_{16}, l_{15}, l_{16}, l_{11}, l_{12}, l_{11}, l_{12})$$
$$V_5 = (l_{15}, l_{15}, l_{16}, l_{16}, l_{27}, l_{28}, l_{27}, l_{28}, l_{27}, l_{28}, l_{27}, l_{28}, l_{15}, l_{16}, l_{15}, l_{16})$$
$$V_6 = (l_{27}, l_{28}, l_{27}, l_{28}, l_{31}, l_{32}, l_{31}, l_{32}, l_{31}, l_{32}, l_{31}, l_{32}, l_{27}, l_{28}, l_{27}, l_{28})$$
$$V_7 = (l_{25}, l_{26}, l_{29}, l_{30}, l_{25}, l_{26}, l_{29}, l_{30}, l_{26}, l_{25}, l_{29}, l_{30}, l_{26}, l_{25}, l_{29}, l_{30})$$
$$V_8 = (l_{21}, l_{22}, l_{22}, l_{21}, l_{23}, l_{23}, l_{24}, l_{24}, l_{24}, l_{24}, l_{23}, l_{23}, l_{21}, l_{21}, l_{22}, l_{22})$$
$$V_9 = (l_{17}, l_{18}, l_{17}, l_{18}, l_{19}, l_{20}, l_{19}, l_{20}, l_{19}, l_{20}, l_{19}, l_{20}, l_{17}, l_{18}, l_{17}, l_{18})$$

where the controlling data block is represented as $V_1 = (l_1, l_2, \ldots, l_{32})$. It is easy to see that this distribution is symmetric relatively to the fifth active layer; therefore, the extension box $\mathbf{E}''$ can be used in the BPI design presented in Figure 4.12 while using the $\mathbf{P}_{32/144}$ box instead of the $\mathbf{P}_{32/128}$ box.

By analogy, we can construct the universal BPI $P^{(V, e_1, e_2)}_{64/352}$ for 64-bit processors using the maximum order permutation network (PN) $\mathbf{P}_{64/352}$, and boxes $T = P^{(e_1)}_{704/1}$ and $T' = P^{(e_2)}_{320/1}$. The hardware implementation cost of different designs of the BPI can be evaluated as a number of the used standard switching elements $\mathbf{P}_{2/1}$ or as a number of NAND gates used for forming the combinational circuit that implements the BPI operation. Evaluation results are presented in Tables 4.7 and 4.8.

All presented variants of the BPI design have relatively low cost of hardware implementation and can be used for embedding in microcontrollers and general-

Table 4.7 Evaluation of the implementation cost for various variants of controlled permutation command (the number of used elementary switches $P_{2/1}$ is indicated)

BPI	Application	Basic PN	T′	T	Total
$P^{(L,e)}_{32/128}$	Cryptographic	128	16	—	144
$P^{(V,e_1,e_2)}_{32/144}$	Multipurpose	144	64	144	352
$P^{(L,e)}_{64/192}$	Cryptographic	192	32	—	224
$P^{(V,e_1,e_2)}_{64/352}$	Multipurpose	352	160	352	864

Table 4.8 Evaluation of the implementation cost in NAND gates

BPI	*Application*	*Basic PN*	T′	T	*Total*
$P^{(L,e)}_{32/128}$	Cryptographic	768	96	—	864
$P^{(V,e_1,e_2)}_{32/144}$	Multipurpose	864	384	864	2112
$P^{(L,e)}_{64/192}$	Cryptographic	1152	192	—	1344
$P^{(V,e_1,e_2)}_{64/352}$	Multipurpose	2112	960	2112	5184

purpose processors. The most interesting variant is the architecture that provides for the universality of the BPI application, i.e., the possibility to perform in one cycle both the variable and arbitrary fixed permutations. This variant seems to be the most attractive one for the manufacturers of processors.

4.5 Hardware Implementation Estimation of the Data-Driven Ciphers

4.5.1 Hardware Implementation Approaches and Architectures

The most common technologies for the encryption algorithms implementation in hardware are the ASIC and FPGA. The application-specific integrated circuit (ASIC) technology guarantees better performance with sufficiently small size. The ASIC devices are based on programmable logic devices (PLDs), gate arrays (GAs), and standard cells (SCs). These devices are tailored to one particular encryption system. ASIC implementation is economical for the market only if such devices are produced in mass quantities. The field programmable gate array (FPGA) technology has several advantages. It is more flexible than the ASIC technology. FPGA implementation is less time-consuming in the development and design phases. The FPGA devices do not require significant initial investment for design, development, and testing. They can be reused for implementing and analyzing many different encryption algorithms and other processes. These devices are commercially available at low prices. The FPGA implementation is economical for the market even if the encryption devices are produced in small quantities.

For all hardware devices there are some critical factors that make the embedding of ciphers in powerful hardware engines a significantly hard process. One such factor is the large number of registers for key storage. RAM blocks are mainly used in hardware implementations; however, the availability of RAM usage is limited. Indeed, the internal memory capacity of many hardware devices is limited, and the use of external RAM reduces the encryption system performance. These factors define the interest in ciphers with simple key scheduling and in those that provide the possibility to generate the round keys on the fly, i.e., directly during the data encryption process (for example, the jth round key can be generated in parallel with performing the $(j - 1)$th encryption round).

Iterated block ciphers are usually implemented using the following two different architectures: (1) loop unrolling architecture (LUA) and (2) pipelined architecture (PA) for both the ASIC and FPGA devices. The LUA architecture used is shown in Figure 4.15a. It is a typical architecture for block cipher implementations because in many applications the block ciphers operate in the cipher block chaining (CBC) mode. According to this architecture, only one data block is transformed at a time. The number of clock cycles (q) required to encrypt a plaintext block or

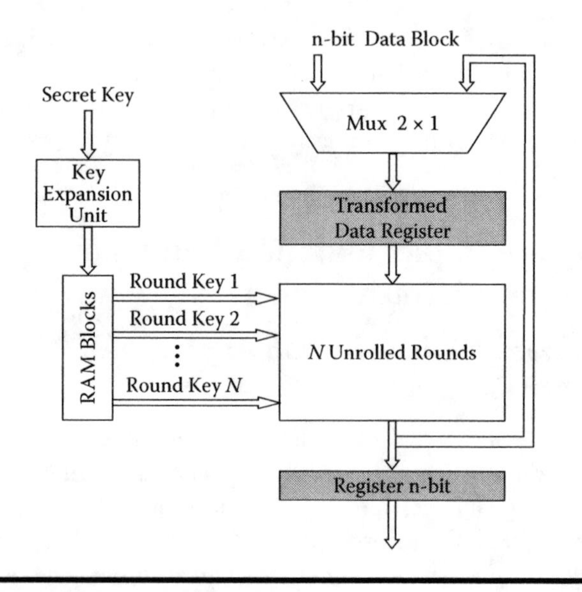

Figure 4.15 Loop unrolling architecture LU-*N*.

decrypt a ciphertext block depends on the number of the specified transformation rounds (R) and the number of the unrolled rounds (N). Usually, N is selected such that the ratio R/N represents an integer, and $q = R/N$. If N rounds are unrolled, then the LUA is denoted as LU-N. The key expansion unit forms the appropriate round keys, which are stored and loaded in the RAM blocks. One round of the encryption algorithm is performed by the data transformation core that represents a flexible combinational logic circuit. This core is supported by the n-bit register and n-bit multiplexer, where $n = 64$ or 128. In the first clock cycle, some n-bit data block is forced into the data transformation core. Then, in each clock cycle, N rounds of the cipher are performed, and the transformed data are stored in the register. According to the LU-N architecture, the n-bit data block is completely transformed every R/N clock cycle.

The PA is a z-stage architecture in which each stage includes R/z unrolled rounds. If z stages are used, then the PA is denoted as P-z. The round keys are stored in the RAM. The intermediate registers are used between the stages to store the transformed data. In the first clock cycle, the first n-bit data block is forced into the first stage. In the second cycle, the first data block passes through R/z transformation rounds and is forced into the second stage. At the same time, the second n-bit data block is forced into the first stage. In the $(z + 1)$th cycle, the first data block passes through the last R/z transformation rounds. At the same time, the $(z + 1)$th n-bit data block is forced into the first stage. Thus, after $z + 1$ clock cycles, the first data block is completely transformed. Then, in each clock cycle, a new data block is completely transformed. We see that z data blocks are transformed simultaneously. The CBC encryption mode cannot be implemented with the P-z architecture for

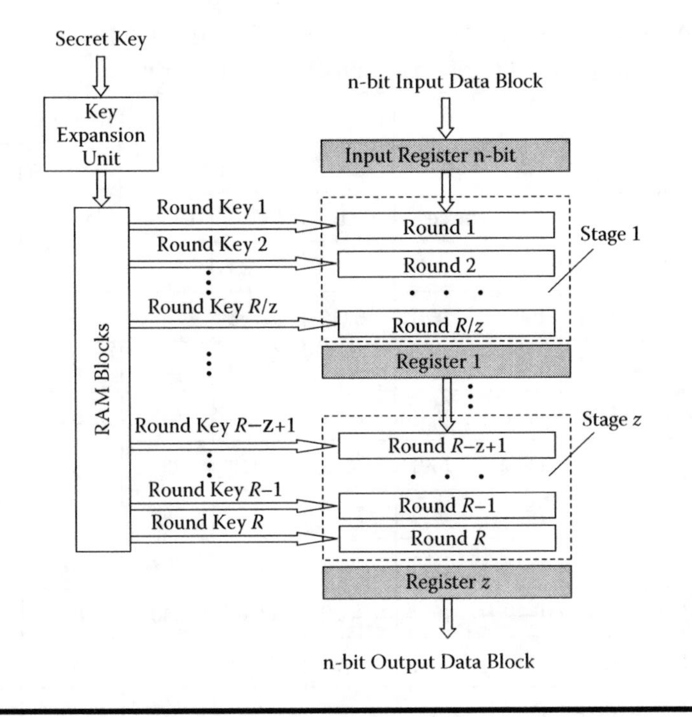

Figure 4.16 Pipeline architecture P-*z*.

$z > 1$; therefore, it is not possible to apply the pipelining in many cryptographic applications. The P-z architecture offers the benefit of high performance and can be used in applications with hard throughput needs (see Figure 4.16). For some given encryption algorithms, the ratio P-z performance/LU-N performance is approximately equal to z. Usually, the LU-N implementation performance is almost the same for all values $N \in \{1, 2, \ldots, R\}$; therefore, the typical LUA corresponds to the value $N = 1$ that provides the cheapest implementation cost. The P-z implementation cost does not significantly depend on z, but the performance is proportional to z. Therefore, the typical PA corresponds to the value $z = R$ that provides the best performance.

In some cases, there is a need for reconfigurable implementations, i.e., to implement several cryptographic algorithms (ciphers and/or hash functions) in a single circuit. The microprocessor design architecture (MDA) can be efficiently applied [66] in such cases. This architecture is designed similar to a typical microprocessor with the RAM memory, control unit, data bus, address bus, and input/output interface. To execute one transformation round, the MDA architecture needs several clock cycles; however, the MDA provides the highest operating frequency for the given device because it has the shortest critical path. The ASIC and FPGA implementations of different cryptographic algorithms are considered in the literature [18–20,29,61,63–66,69–71].

Table 4.9 Implementation synthesis results of some DDP-based encryption algorithms (Xilinx Vitrex FPGA device)

Encryption algorithm and reference	Architecture and reference	FPGA device (Xilinx Vitrex)				
		Covered area			*F* (MHz)	Rate (Mbps)
		CLBs	FGs[a]	DFFs[b]		
CIKS-1 [38]	LU-1 [64]	907	1723	192	81	648
CIKS-1 [38]	P-8 [64]	6346	14128	576	81	5184
SPECTR–H64 [13]	LU-1 [64]	713	1320	203	83	443
SPECTR–H64 [13]	P-12 [64]	7021	10456	832	83	5312
DDP-64 [49]	LU-1 [49]	615	1230	207	85	544
DDP-64 [49]	P-10 [49]	3440	6880	640	95	6100
Cobra–H64 [65]	LU-1 [65]	615	1229	204	82	525
Cobra–H64 [65]	P-10 [65]	3020	6040	640	85	5500
Cobra–H128 [65]	LU-1 [65]	2364	4728	399	86	917
Cobra–H128 [65]	P-12 [65]	22080	44160	?	90	11500

[a] Functional generators.
[b] D Flip-Flops.

4.5.2 Hardware Implementations of the DDP-Based Ciphers

A number of hardware implementations of the DDP-based ciphers are proposed in several studies [49,64,65]. It has been shown that DDP operations can be efficiently implemented in both the FPGA and ASIC types of devices. The implementation synthesis results are presented in Tables 4.9 and 4.10.

4.5.3 Hardware Implementations of the DDO-Based Ciphers

The ASIC and FPGA implementation results of a number of DDO-based ciphers are presented in the literature [39,44,46,50], which show that such ciphers are the most efficient for hardware implementation against the known block encryption algorithms. These ciphers use the CSPNs as main primitives in their design. For the given topology, the FPGA implementation cost of CSPNs constructed using the $\mathbf{F}_{2/1}$ and $\mathbf{F}_{2/2}$ elements is equal to the implementation cost of the PNs. However, the DDOs are more efficient as cryptographic primitive than DDPs for the given topology; therefore, using DDOs we can design secure block ciphers that have significantly shorter critical path

Table 4.10 Implementation synthesis results of some DDP-based ciphers (0.33-μm ASIC technology)

Encryption algorithm and reference		Implementation architecture and reference		Area (sq mil)	*F* (MHz)	Rate (Mbps)
CIKS-1	[38]	LU-1	[64]	3,456	93	744
CIKS-1	[38]	P-8	[64]	21,036	95	5,824
SPECTR–H64	[13]	LU-1	[64]	3,194	91	485
SPECTR–H64	[13]	P-12	[64]	32,123	94	6,016
DDP-64	[49]	LU-1	[49]	2,620	92	589
DDP-64	[49]	P-10	[49]	14,050	101	6,500
Cobra–H64	[65]	LU-1	[65]	2,694	100	640
Cobra–H64	[65]	P-10	[65]	14,640	110	7,100
Cobra–H128	[65]	LU-1	[65]	6,364	90	1,000
Cobra–H128	[65]	P-12	[65]	48,252	95	12,100

than the DDP-based algorithms. For example, see Section 3.5.3 in Chapter 3, which describes the construction of new ciphers from the DDP-based ones by replacing the DDP boxes with the DDO boxes having the same topology. By such replacement, the number of the encryption rounds has been significantly reduced. Thus, for the FPGA hardware implementation, the DDO boxes are preferable to DDP boxes.

In case of the ASIC devices, the implementation costs of the CSPNs and PNs are different because, in general, the implementation cost of the $\mathbf{P}_{2/1}$, $\mathbf{F}_{2/1}$, and $\mathbf{F}_{2/2}$ elements are different. To estimate the ASIC implementation costs of different CSPNs, Moldovyan and others [50] present the results given in Table 4.11. We see that the fastest and the cheapest elementary controlled involutions $\mathbf{F}_{2/1}$ are CEs #3 and #4, i.e., elements (e,g) and (e,h). The elements #12, #17, #2, #5, #7, and #13 are also attractive for block cipher design. Each of the elements #3 and #4 can be implemented using area 2 sq mil and they operate with frequency 1.923 GHz. The switching element $\mathbf{P}_{2/1}$ coves area 3 sq mil and performs with frequency 2.12 GHz [50].

The FPGA and ASIC implementation synthesis results of the number of the DDO-based ciphers are presented in Tables 4.12 and 4.13 in comparison with the implementation results of some well-known encryption algorithms such as AES, Serpent, Twofish, RC6, IDEA, and 3DES. In comparison with these algorithms and the DDP-based ciphers, the DDO-based algorithms are significantly faster for the LU-1 implementation architecture and significantly cheaper for the P-*R*

Table 4.11 The ASIC implementation synthesis results for all nonlinear elementary controlled involutions

No.	Type	CE	BF		ASIC 0.33 μm	
			$f_1(x_1,x_2,v)$	$f_2(x_1,x_2,v)$	Area (sq mil)	*F* (GHz)
1	**R**	(g,e)	$x_1v\oplus x_2v\oplus x_1$	$x_2v\oplus x_1\oplus x_2$	3	0.952
2	**Q**	(g,h)	$x_2v\oplus x_1$	$x_1v\oplus x_1\oplus x_2$	3	1.370
3	**R**	(e,g)	$x_1v\oplus x_2v\oplus x_2$	$x_2v\oplus x_1$	2	1.923
4	**R**	(e,h)	$x_1v\oplus x_2$	$x_1v\oplus x_2v\oplus x_1$	2	1.923
5	**Q**	(h,g)	$x_2v\oplus x_1\oplus x_2$	$x_1v\oplus x_2$	3	1.351
6	**R**	(h,e)	$x_1v\oplus x_1\oplus x_2$	$x_1v\oplus x_2v\oplus x_2$	3	0.952
7	**Q**	(g,i)	$x_2v\oplus x_1\oplus v$	$x_1v\oplus x_1\oplus x_2$	3	1.370
8	**R**	(g,f)	$x_1v\oplus x_2v\oplus x_1\oplus v$	$x_2v\oplus x_2\oplus x_1\oplus v$	4	0.833
9	**Q**	(i,g)	$x_2v\oplus x_2\oplus x_1\oplus v\oplus 1$	$x_1v\oplus x_2$	4	0.735
10	**R**	(f,g)	$x_1v\oplus x_2v\oplus x_2\oplus v\oplus 1$	$x_2v\oplus x_1\oplus v\oplus 1$	3	0.885
11	**R**	(i,f)	$x_1v\oplus x_2\oplus x_1\oplus 1$	$x_1v\oplus x_2v\oplus x_2\oplus v$	3	0.885
12	**R**	(f,i)	$x_1v\oplus x_2\oplus 1$	$x_1v\oplus x_2v\oplus x_1\oplus v\oplus 1$	2	1.724
13	**Q**	(h,j)	$x_2v\oplus x_1\oplus x_2$	$x_1v\oplus x_2\oplus v$	3	1.351
14	**Q**	(j,h)	$x_2v\oplus x_1$	$x_1v\oplus x_2\oplus x_1\oplus v\oplus 1$	3	0.585
15	**R**	(j,f)	$x_1v\oplus x_2v\oplus x_1\oplus v$	$x_2v\oplus x_2\oplus x_1\oplus 1$	3	1.282
16	**R**	(f,h)	$x_1v\oplus x_2\oplus v\oplus 1$	$x_1v\oplus x_2v\oplus x_1\oplus v\oplus 1$	3	0.885
17	**R**	(f,j)	$x_1v\oplus x_2v\oplus x_2\oplus v\oplus 1$	$x_2v\oplus x_1\oplus 1$	2	1.724
18	**R**	(e,j)	$x_1v\oplus x_2v\oplus x_2$	$x_2v\oplus x_1\oplus v$	4	0.952
19	**R**	(j,e)	$x_1v\oplus x_2v\oplus x_1$	$x_2v\oplus x_2\oplus x_1\oplus v\oplus 1$	4	0.599
20	**Q**	(j,i)	$x_2v\oplus x_1\oplus v$	$x_1v\oplus x_2\oplus x_1\oplus v\oplus 1$	4	0.741
21	**R**	(i,e)	$x_1v\oplus x_2\oplus x_1\oplus v\oplus 1$	$x_1v\oplus x_2v\oplus x_2$	4	0.599
22	**Q**	(i,j)	$x_2v\oplus x_2\oplus x_1\oplus v\oplus 1$	$x_1v\oplus x_2\oplus v$	4	0.629
23	**R**	(h,f)	$x_1v\oplus x_1\oplus x_2\oplus v$	$x_1v\oplus x_2v\oplus x_2\oplus v$	4	0.833
24	**R**	(e,i)	$x_1v\oplus x_2\oplus v$	$x_1v\oplus x_2v\oplus x_1$	3	0.885

Table 4.12 Implementation synthesis results of some block encryption algorithms (FPGA device)

Cipher	*R* (number of rounds)	Architecture		Area (CLBs)	*F* (MHz)	Data rate (Mbps)
DDO-64	8	LU-1	[40]	350	130	924
Eagle-64	8	LU-1	[44]	305	142	1,050
CHESS-64	8	LU-1	[50]	665	91.5	732
Eagle-128	10	LU-1	[46]	781	92	1,177
Eagle-128	10	P-10	[46]	4,120	95	12,160
AES	10	LU-1	[64]	2,358	22	259
AES	10	LU-1	[12]	3,528	25.3	294
AES	10	LU-2	[12]	5,302	14.1	300
AES	10	LU-1	[10]	3,552	54	493
AES	10	P-10	[63]	17,314	28.5	3.650
Serpent	32	LU-8	[12]	7,964	13.9	444
RC6	20	LU-1	[12]	2,638	13.8	88.5
Twofish	16	LU-1	[12]	2,666	13	104
IDEA	8	LU-1	[9]	2,878	150	600
3DES	3×16	?	[55]	604	165	587

implementation architecture. For both the implementation technologies ASIC and FPGA, the DDO-based ciphers DDO-64, Eagle-64, and Eagle-128 are very attractive due to their high performance and low cost. Besides, the DDO-based ciphers require few block RAMs and, therefore, they are efficient for the variety of FPGA device types.

4.5.4 Hardware Implementation Estimations of the Ciphers Based on Switchable DDOs

The FPGA implementation estimations of the iterative ciphers SCO-1, SCO-2, and SCO-3 [41] based on SCO have been performed by Sklavos and Koufopavlou [66] using the LU-1 and MDA architectures. The MDA used is shown in Figure 4.17. The results obtained are shown in Tables 4.14 and 4.15.

Table 4.13 ASIC implementation synthesis results of some block ciphers

Cipher	Number of rounds	Architecture		Technology, μm	Area (Gates)	*F* (MHz)	Data rate (Mbps)
Eagle-128	10	LU-1	[46]	0.33	3,104 sq mil	110	1,408
Eagle-128	10	P-10	[46]	0.33	16,780 sq mil	112	14,336
AES	10	?	[59]	?	256,000	32	7,500
AES	10	LU-10	[15]	0.35	612,834	15.2	1,950
Serpent	32	LU-32	[15]	0.35	503,770	7.28	932
RC6	20	LU-20	[15]	0.35	1,643,037	1.59	204
Twofish	16	LU-16	[15]	0.35	431,857	3.08	394
3DES	3×16	LU-48	[15]	0.35	148,147	6.4	407
DES	16	LU-16	[15]	0.35	54,405	18.1	1,161

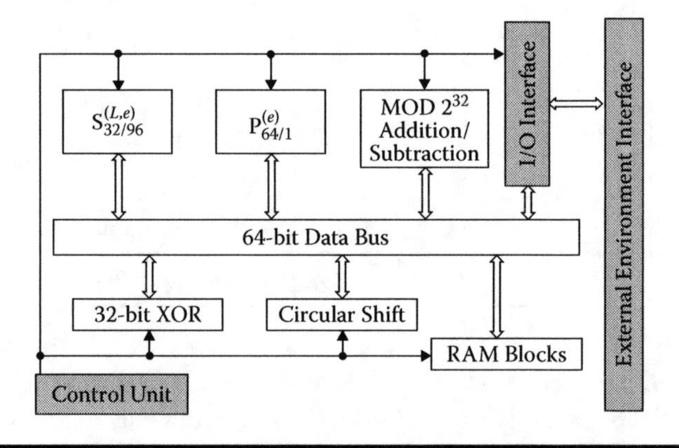

Figure 4.17 Reconfigurable MDA architecture [66] used for implementation of three algorithms SCO-1, SCO-2, and SCO-3.

4.5.5 Hardware Implementation Efficacy Comparison

To compare hardware implementation efficacy of different ciphers, the performance/cost model is usually used. Let us now additionally use the performance/(cost·frequency) model. Using these two comparison models, the different ciphers are compared in Tables 4.16 and 4.17, and in Figure 4.18. Note that Serpent is implemented by Elbirt and others [12] using the LU-8 architecture, which is less efficient relative to the first comparison model than the LU-1 architecture,

Table 4.14 Implementation synthesis results of the SCO-based ciphers

		Xilinx FPGA Vitrex device v200pq240				
		Covered area			*F*	Rate
Cipher	R	CLBs	FGs	DFFs	(MHz)	(Mbps)
SCO-1	8	860	1,720	352	58	464
SCO-2	12	422	844	320	76	405
SCO-3	12	288	576	320	82	438
Hawk-64[a]	8	470	960	320	82	650

[a] Our estimation.

Source: From Sklavos, N. and Koufopavlou, O. 2003. Architectures and FPGA Implementations of the SCO (−1, −2, −3) Ciphers Family. Proceedings of the 12th International Conference on Very Large Scale Integration, (IFIP VLSI SOC '03). Darmstadt, Germany. pp. 68–73. With permission.

Table 4.15 Reconfigurable implementation results of the SCO-based ciphers

	Xilinx FPGA Vitrex device v200pq240						
	Covered area			*F*			Rate
Architecture	CLBs	FGs	DFFs	(MHz)	Algorithm (operation)	Number of clock cycles	(Mbps)
LU-1	975	2,050	352	47	SCO-1	8	464
					SCO-2	12	405
					SCO-3	12	438
MDA	890	1,780	372	307	SCO-1	8×13	188
					SCO-2	12×6	272
					SCO-3	12×8	204

Source: From Sklavos, N. and Koufopavlou, O. 2003. Architectures and FPGA Implementations of the SCO (−1, −2, −3) Ciphers Family. Proceedings of the 12th International Conference on Very Large Scale Integration, (IFIP VLSI SOC '03). Darmstadt, Germany. pp. 68–73. With permission.

approximately by a factor 8. Therefore, we should estimate the efficacy of Serpent as 56 (see Table 4.16) to ≈450 Kbps/#CLBs. Similarly, for the first comparison model, the integral efficiency value (E_{mod1}) of the AES, Serpent, RC6, Twofish, Triple-DES, and DES algorithms implemented in the LU-N architecture where $N = R$,

Table 4.16 Integral efficacy estimation results of the FPGA implementation

Cipher	Architecture	Block size, bit	*R*	Integral efficacy	
				Kbps/number of CLBs	Kbps/(number of CLBs·GHz)
Cobra-H64	LU-1 [65]	64	10	850	10,400
Cobra-H64	P-10 [65]	64	10	1,830	21,400
Cobra-H128	LU-1 [65]	128	12	390	4,500
Cobra-H128	P-12 [65]	128	12	520	5,800
DDP-64	LU-1 [49]	64	10	880	10,600
DDO-64	LU-1 [39]	64	8	2,640	20,200
Eagle-64	LU-1 [44]	64	10	3,440	24,200
CHESS-64	LU-1 [50]	64	8	1,100	12,000
Eagle-128	LU-1 [46]	128	10	1,510	16,400
Eagle-128	P-10 [46]	128	10	2,900	30,500
AES	LU-1 [64]	128	10	110	5,000
AES	LU-1 [12]	128	10	83	3,300
AES	LU-2 [12]	128	10	57	4,400
AES	LU-1 [10]	128	10	138	2,560
AES	P-10 [63]	128	10	210	7,400
Serpent	LU-8 [12]	128	32	56	4,000
RC6	LU-1 [12]	128	20	34	2,400
Twofish	LU-1 [12]	128	16	39	3,000
IDEA	LU-1 [9]	128	8	34	2,400
3DES	? [55]	64	3×16	940	5,700
SCO-2	LU-1 [41]	64	12	960	12,600
SCO-3	LU-1 [41]	64	12	1,510	18,400
Hawk-64[a]	LU-1 [41]	64	8	1,380	16,900

[a] Our estimation

Table 4.17 Integral efficacy estimation results of the ASIC implementation

Cipher	Architecture		Block size, bit	*R*	Integral efficacy	
					Kbps/number of Gates	Kbps/(number of Gates·GHz)
SPECTR–H64	LU-1	[64]	64	12	75	820
SPECTR–H64	P-10	[64]	64	12	94	1,000
Cobra-H64	LU-1	[65]	64	10	118	1,180
Cobra-H64	P-10	[65]	64	10	247	2,220
Cobra-H128	LU-1	[65]	128	12	78	867
Cobra-H128	P-12	[65]	128	12	125	1,310
DDP-64	LU-1	[49]	64	10	112	1,210
DDP-64	P-10	[49]	64	10	232	2,300
Eagle-128	LU-1	[46]	128	10	227	2,060
Eagle-128	P-10	[46]	128	10	425	3,800
AES	?	[59]	128	10	15	469
AES	LU-10	[12]	128	10	3.2	209
Serpent	LU-32	[12]	128	32	1.85	253
RC6	LU-20	[12]	128	20	0.124	78
Twofish	LU-16	[12]	128	16	0.913	296
Triple-DES	LU-48	[55]	64	3x16	2.75	1,180
DES	LU-16	[9]	64	16	21.34	430

depends on the value N, i.e., $E_{\text{mod1}} = E_{\text{mod1}}(N)$. We can roughly estimate $E_{\text{mod1}}(1) < N{\cdot}E_{\text{mod1}}(N)$. In Figure 4.19*a*, we present the values $E_{\text{mod1}}(N)$ and $E_{\text{mod1}}(1)$ calculated using the last formula. (For the second comparison model, the integral efficiency value (E_{mod2}) of the encryption algorithms implemented in the LU-N architecture is almost independent of the value N.) Comparison results show that the data-driven ciphers have significantly higher values for the integral efficacy of the hardware implementation than other known ciphers, except the DES algorithm, for which an approximate estimation gives $E_{\text{mod1}}(1) < 341$ *Kbps/#Gates*. We presume that the Eagle-64 algorithm has the efficiency $E_{\text{mod1}}(1) \approx 270$ *Kbps/#Gates*, but this should, however, be verified by the ASIC implementation simulation.

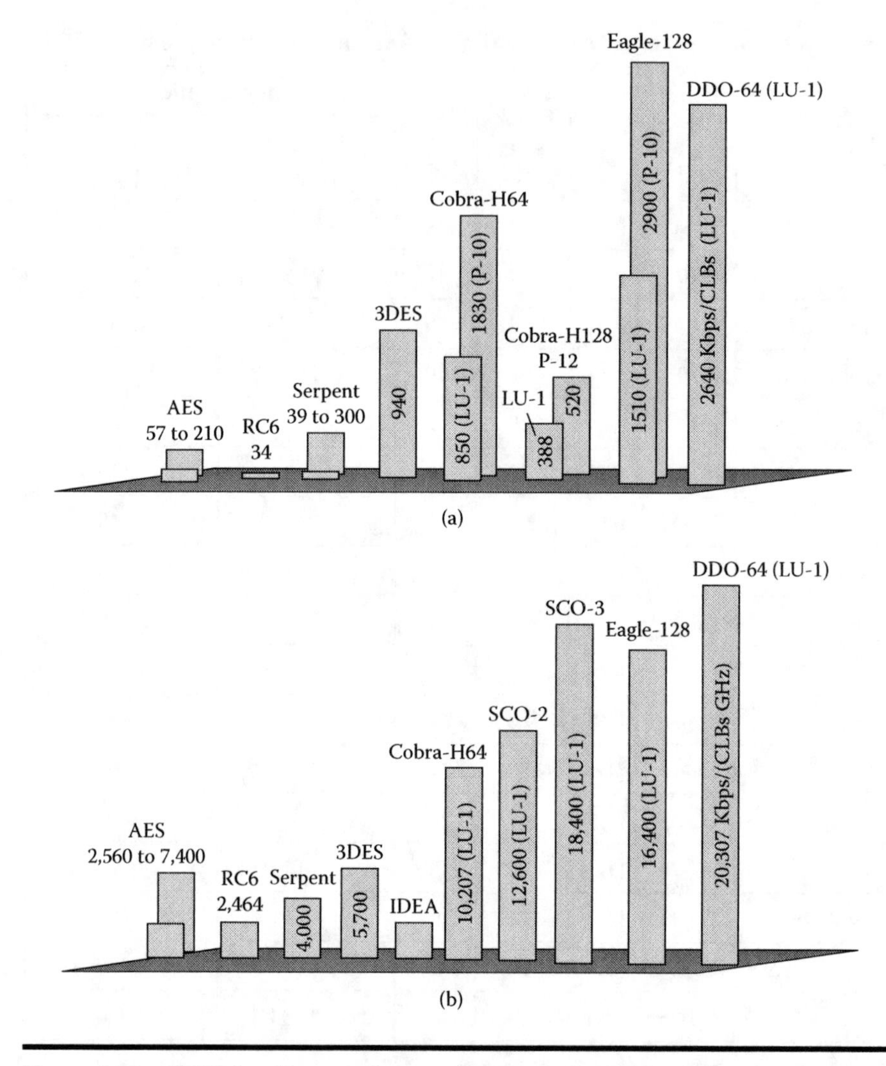

Figure 4.18 FPGA implementation results comparison: (*a*) performance/area, Kbps/CLBs; (*b*) performance/(area·frequency), Kbps/(CLBs·GHz).

4.6 On-Fly Expansion of the Secret Key

The previously considered block ciphers based on DDOs use simple key scheduling, although conventional ciphers usually use sufficiently complex key scheduling (an exception is the Russian standard GOST [60]). Key expansion is a simple and efficient way to prevent different theoretical weaknesses of the block ciphers. In any case, using a complex key scheduling instead of a simple one is reasonable from the security point of view. However, to perform the key expansion, we need additional hardware resources (for hardware implementation) or precalculation of all round

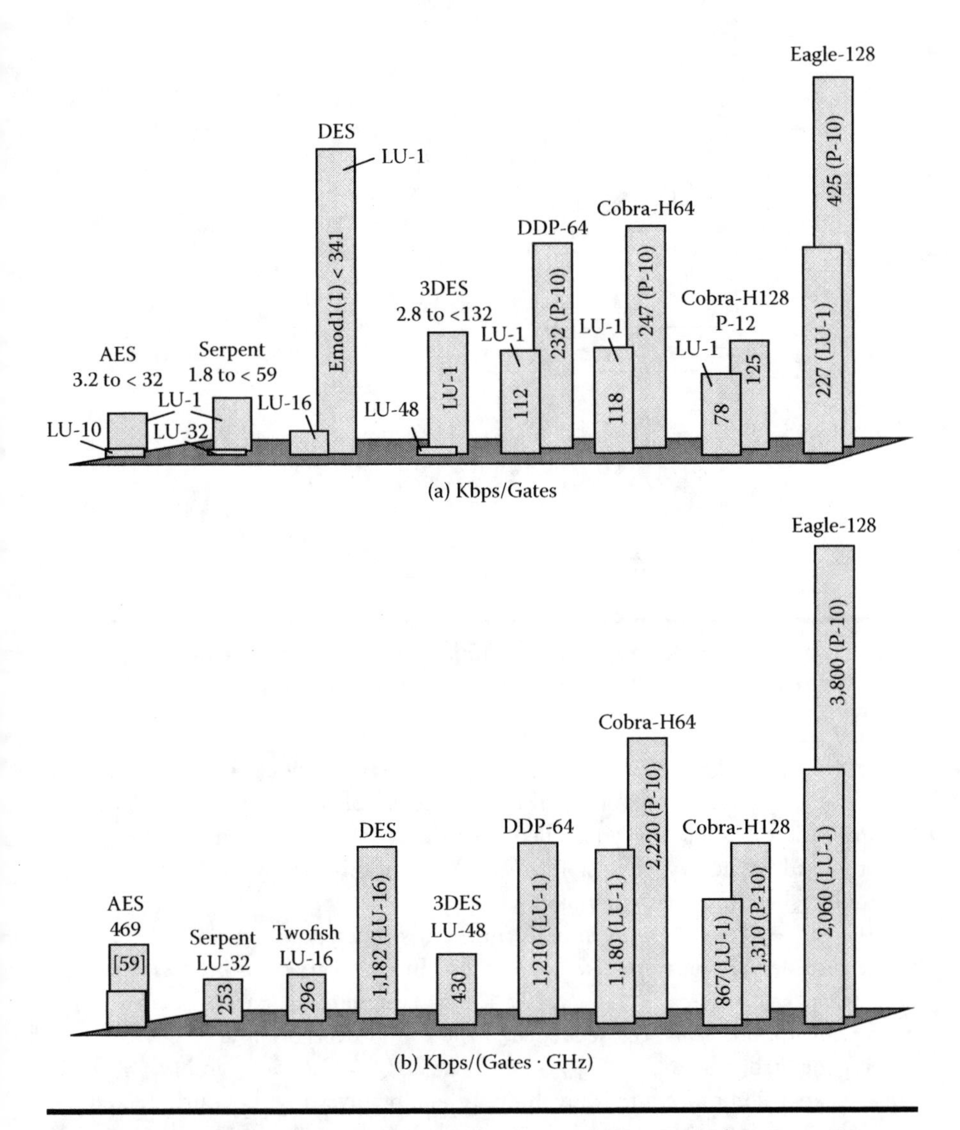

Figure 4.19 ASIC implementation results comparison: (*a*) Performance/Area, *Kbps/Gates*; (*b*) Performance/(Area·Frequency), *Kbps/(Gates·GHz)*.

keys. The latter reduces the encryption performance in case of frequent change of keys. In this section, we consider the technique of generating the round keys on the fly. The generalized scheme of such a technique is described as follows: The first round key Q_1 is a part of the secret key. During the first encryption round, the second round key Q_2 is computed. Then the jth round key Q_j, where $j = 3, 4, \ldots, R + 1$, is computed in parallel with performing the $(j - 1)$th encryption round (we consider a cipher with final transformation; the Q_{R+1} key is used to perform

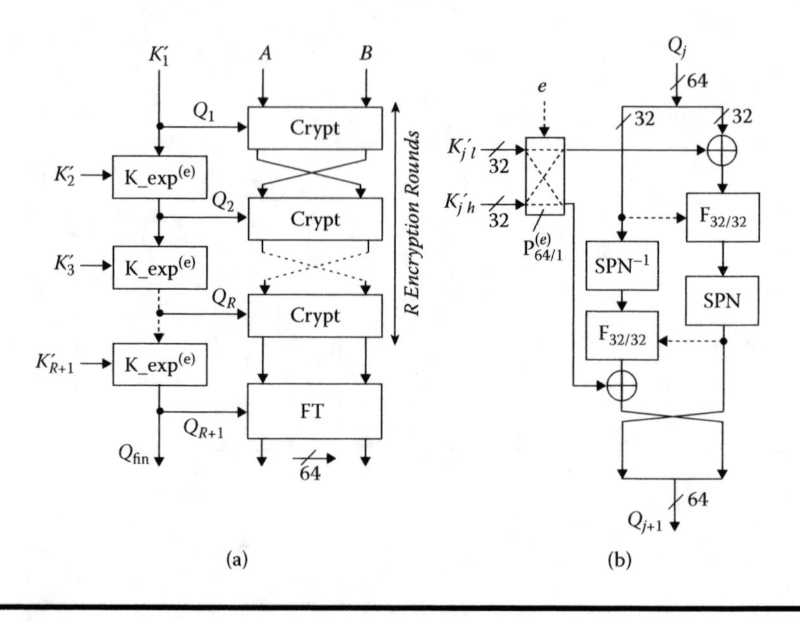

Figure 4.20 Iterative encryption algorithm with the on-fly key expansion (*a*) and structure of the key expansion procedure iteration K_exp$^{(e)}$ (*b*).

the final transformation). The last round key Q_{fin} is saved. To implement efficiently the on-fly key expansion, we should use some iterative key expansion procedure, the latency of which is less or equal to the latency of the encryption round. The key expansion procedure is controlled with the secret key, the parts of which are directly used in the key expansion procedure iterations (i.e., we have some source secret key scheduling that is simple).

The $Q_{fin} = Q'_1$ key is used to perform the first decryption round and compute the second decryption round key and so on. In the decryption process, the Q'_{j+1} decryption key, where $j = 1, 2, \ldots, R + 1$, is computed in parallel with the jth decryption round. Thus, the sequence of the round decryption keys $Q'_1, Q'_2, \ldots, Q'_{fin}$ is generated, where $Q'_1 = Q_{fin}$, $Q'_2 = Q_{R+1}, \ldots, Q'_j = Q_{R-j+2}, \ldots$, and $Q'_{fin} = Q_1$. If the key expansion procedure is involution, then, by inverting the source secret key scheduling, we invert the expanded key scheduling.

The next efficient variant of implementing the on-fly secret key expansion is connected with the use of the switchable key expansion procedure iterations. The last variant is illustrated in Figure 4.20. The key expansion procedure iteration **K_exp**$^{(e)}$ includes the same operations $\mathbf{F}_{32/32}$, SPN, and SPN^{-1} as the right branch of the Eagle-128 cipher. The **K_exp**$^{(e)}$ function is not involution, but it is a switchable function, i.e., we have

$$\mathbf{K\_exp}^{(0)} = (\mathbf{K\_exp}^{(1)})^{-1} \tag{4.31}$$

Table 4.18 Two possible variants of the secret-key scheduling

Variant		$i =$	1	2	3	4	5	6	7	8	9	10
1	$K'_i =$	(e = 0)	K_1	K_2	K_3	K_4	K_1	K_4	K_1	K_4	K_3	K_2
		(e = 1)	Q_{fin}	K_2	K_3	K_4	K_1	K_4	K_1	K_4	K_3	K_2
2	$K'_i =$	(e = 0)	K_1	K_2	K_3	K_4	K_1	K_3	K_4	K_2	K_4	K_3
		(e = 1)	Q_{fin}	K_3	K_4	K_2	K_4	K_3	K_1	K_4	K_3	K_2

The switchability property of the $\mathbf{K\_exp}^{(e)}$ procedure is due to the use of the transposition box $\mathbf{P}^{(e)}_{64/1}$. The considered variant of the key expansion can be included in the Eagle-128 algorithm with the 256-bit secret key $K = (K_1, K_2, K_3, K_4)$. Table 4.18 presents two possible variants of the secret-key scheduling. The first variant of the secret-key scheduling is a bit more attractive because, at steps having numbers $i = 2, 3, \ldots, 10$, the used elements of the secret key are the same for both the data encryption and the data decryption procedures; therefore, it is a bit cheaper for hardware implementation. For the first variant in the decryption process, the scheduling of round keys is reversed due to switching the operation $\mathbf{K\_exp}^{(e)}$ and using Q_{fin} instead of K_1 at the first step of the key processing. For the second variant, the secret-key scheduling is different for encryption and for decryption.

The FPGA implementation of such key expansion algorithms requires about 160 additional configurable logic blocks (CLBs). If it is included in the Eagle-128 algorithm, the integral efficiency of Eagle-128 implemented in the LU-1 (P-10) architecture becomes equal to the values 1,250 *Kbps/#CLBs* and 13,600 *Kbps/(#CLBs·GHz)* (2,126 *Kbps/#CLBs* and 22,380 *Kbps/(#CLBs·GHz)*), correspondingly to the first and the second comparison models. Thus, including the on-fly key expansion procedure in the Eagle-128 algorithm reduces its FPGA implementation efficacy for the LU-1 (P-10) architecture to 83% (73%) of the integral efficacy value of Eagle-128 with simple key scheduling. However, the modified version of Eagle-128 remains significantly more efficient in hardware than AES, Serpent, RC6, Twofish, and Triple-DES.

4.7 Conclusions

Results of this chapter can be summarized as follows:

- The CSPNs constructed using CEs that are involutions, and having recursive topologies of different orders ($h = 1, 2, \ldots, n/4, n$), implement the sets of pairs of the mutually inverse transformation modifications.
- The CSPNs can be efficiently used to construct reversible SDDO boxes.

- Using the CSPNs with symmetric topology and symmetric distribution of the bits corresponding to two halves of the controlling data subblock, it is possible to reduce the hardware implementation cost of the SDDO to the value ≈117% of the cost of the analogous ordinary (i.e., nonswitchable) DDO box.
- The switchable CSPNs of the orders $h = 1, 2, \ldots, n/4$ can be constructed using $s = \log_2 nh$ active layers. The switchable CSPNs of the order $h = n$ can be constructed using $s = 2 \cdot \log_2 n - 1$ active layers.
- Two different switching mechanisms providing reversibility of the DDO boxes have been presented: (1) transposing the controlling bits corresponding to the CEs in symmetric positions and (2) inverting the controlling bits. In the second case, the bit controlling the ith CE in the jth active layer should also control the ith CE in the $(s - j + 1)$th active layer, where s is the number of active layers in the DDO box.
- The architecture of the universal controlled bit permutation instruction has been proposed for embedding in the general purpose processors. The proposed instruction allows to perform in one cycle both the variable permutations (i.e., DDP) and the arbitrary, given fixed permutations.
- Several ciphers based on SDDOs and suitable for cheap FPGA and ASIC implementation have been described.
- Hardware implementation estimations of different DDP-based, DDO-based, and SDDO-based ciphers have been presented and compared with the implementation results of the AES, Serpent, RC6, Twofish, Triple-DES, and some other block ciphers.

Chapter 5

Data-Driven Ciphers Suitable for Software Implementation

This chapter summarizes the results of studies [32–37] considering different variants of the fast software-oriented block ciphers based on data-dependent subkey selection (DDSS). Two different approaches to using the DDSS are proposed. In the first approach, the selected subkeys are directly combined with transformed data subblocks, and in the second one, the selected subkeys are used to modify the intermediate key variables that are then combined with the data subblocks. The ciphers based on the DDSS are efficient only in software because they use significantly large expanded key that should be precomputed and stored.

5.1 A Class of Ciphers Based on Data-Dependent Subkey Selection

There are two types of data-driven operations efficient on the general purpose microprocessors available today: data-dependent rotation (DDR) and DDSS. The DDRs are extensively used in the well-known ciphers RC5 and RC6. Ciphering mechanism described in the following text characterizes a class of block encryption functions based on DDSS. We suppose that an expanded encryption key that represents a sequence containing 2^δ b-bit subkeys $Q[j]$, where $j = 0, 1, 2, \ldots, 2^\delta - 1$, is used. This sequence is to be precomputed, depending on the secret key, the length of which is 128 to 256 bits. The input data block is represented as a concatenation of b-bit subblocks:

$$M = (B^{(0)}, B^{(1)}, \ldots, B^{(w-1)}) \quad (5.1)$$

Encryption and decryption are executed as a sequence of elementary conversion steps:

$$C = E(M) = e_k(e_{k-1}(\ldots e_2(e_1(M)))) \quad (5.2)$$

$$M = D(C) = e'_k(e'_{k-1}(\ldots e'_2(e'_1(C)))) \quad (5.3)$$

where $e_1, e_2, \ldots, e_k$ and $e'_1, e'_2, \ldots, e'_k$ are elementary encryption and decryption functions; C is a cipher text block. Elementary decryption function e'_i is the inverse of function e_{k-i+1}. Elementary encryption functions have the following structures:

$$j(s, g') = (B^{(g')})^{<d(s-1)<} \bmod 2^\delta \quad (5.4)$$

$$B^{(g)} \leftarrow \mathrm{e}_i(B^{(g)}, Q[j(s,g')]) \quad (5.5)$$

where $B^{<x<}$ denotes to-left rotation of B by x bits, $g' = (i - 1) \bmod w$, $g = i \bmod w$, d is an integer ($d | b$ and $1 < d < b$), i is a counter of elementary encryption steps $i = 1, 2, \ldots, k$ ($k \geq wb/d$), and s is a counter of encryption subrounds $s = (i - 1) \bmod (b/d) + 1$ (where $i = 1, 2, \ldots, b/d$). Integer d defines the number of encryption subrounds in one round. One encryption round includes wb/d consequent elementary steps. The value $k = wb/d$ corresponds to the case when all binary digits of the input data block M will be used to control the subkey selection. Thus, in one round of the encryption function under consideration, each input bit influences the subkey selection. However, it may happen that two different input data blocks define the same set of selected subkeys. If this happens, it is a weakness from a general cryptanalytic point of view. To define the dependence of subkey selection on every input bit, one must impose certain conditions on the used elementary encryption functions.

We shall consider elementary encryption functions e_i which, for fixed $B^{(g)}$, define a permutation $\gamma(Q[j])$ of a set of numbers $\{0, 1, 2, \ldots, 2^b - 1\}$. It appears acceptable to use the same function e for all conversion steps as well as to define different elementary functions. The equation 5.4 defines the selection of a subkey according to the current value of the subblock $B^{(g')}$ and the number of the current subround. The selected subkey is used for conversion of the subblock $B^{(g)}$ where $g \neq g'$. Elementary decryption functions have the analogous structure:

$$j(s,q') = (B^{(q')})^{<b-sd<} \bmod 2^\delta \quad (5.6)$$

$$B^{(q)} \leftarrow e'_i(B^{(q)}, Q[j(s,q')]) \quad (5.7)$$

where $q = w + 1 - i \bmod w$, $q' = w - i \bmod w$, and $s = (i - 1) \bmod (b/d) + 1$. Due to such a mechanism of subkey selection, for arbitrary encryption key $\{Q[j]\}$ the same subkeys are used at the respective decryption and encryption steps.

Transformation of some given block M can be described by the following generalized formula: $C = f(M, K)$, where $K = (Q[z_1], Q[z_2], \ldots, Q[z_k])$ is a (kb)-bit binary vector, and $Q[z_i]$, $i = 1, 2, \ldots, k$ are selected subkeys. Selection of the ordered set $\{Q[z_i]\}$ depends on both the input message and the encryption key. One can say that the given input block M generates an ordered set of indices $\{z_i\}$ which can be put into the correspondence with value $Z = \sum_{i=1}^{k} 2^{8(i-1)} z_i$. Vector K is a set of subkeys used to encrypt the given data block M. There are $N_m = 2^{bw}$ different input data blocks, but for a given expanded key, the number of different vectors K is equal to $N_k \approx (2^{\delta wb/d})^R$, where R is the number of assigned conversion rounds. Because the subkeys are selected pseudorandomly, and $N_k/N_m = 2^{wb(\delta R/d-1)}$, it is reasonable to presume that, for $\delta R/d - 1 \geq 1.5$ and $wb \geq 64$ bits, the probability of generating two equal vectors K (or values Z) for two different input blocks is very low. Nevertheless, a class of encryption functions for which the subkey sets are provably unique for all possible input data blocks appears to be preferable for application in the block cipher design based on the considered mechanism of the DDSS.

Let us impose an additional condition on elementary ciphering functions and consider the converted subblocks as a concatenation of the most significant (H) and the least significant (L) vectors:

$$B = (L, H) = H \cdot 2^{d(s'-1)} + L \tag{5.8}$$

where the size of the L binary vector depends on the number of the current conversion subround $l = d(s' - 1)$ (where $s' = s$ for $i \neq w, 2w, \ldots, wb/d$ or, otherwise, $s' = s + 1$). The size of the vector H also changes corresponding from one subround to another. The following statement is quite evident:

Statement 5.1. *If all elementary encryption functions e_i define a permutation of the value of the most significant subblocks H, $k \geq wb/d$, and $d \leq \delta$, then for all input data blocks M the unique ordered sets of indices $\{z_i\}$, $i = 1, 2, \ldots, k$ are generated.*

The condition formulated in Statement 5.1 means that elementary encryption functions should have the following structure:

$$B \leftarrow e(B) = \beta(H) \cdot 2^{d(s'-1)} + f(L) \tag{5.9}$$

where f is a function satisfying condition $f(L) \leq 2^{d(s'-1)} - 1$, and $\beta(H)$ is a permutation of the set of numbers $\{0, 1, 2, \ldots, 2^{b-d(s'-1)} - 1\}$. For such class of resultant encryption functions, one can formulate the following theorem:

Theorem 5.1. *If $k = wb/d$ and $d = \delta$, then for arbitrary encryption key $\{Q_j\}$ and arbitrary ordered set of indices $\{z_i\}$, where $i = 1, 2, \ldots, k$, there exists the unique input data block M, which generates the given set of indices.*

Proof. The cardinal number of the set of possible values Z is $\#Z = (2^{\delta})^k$. If $d = \delta$ and $k = wb/d$, then $\#Z = 2^{wb} = \#M$. Taking into account Statement 5.1, we obtain the one-to-one correspondence for the sets of possible values Z and M. This means that for arbitrary set $\{z_i\}$ there exists the unique M which defines the generation of a given set of indices.

Thus, we have defined a class of ciphers with unique subkey selection for all input data blocks. Algorithm realization of such ciphers is described in Section 5.3. The fact that subkey selection is not predetermined appears to be the main factor contributing significantly to the security of the ciphers. This can be used for creation of the flexible ciphers with the key-dependent encryption algorithm modifications, which include the DDSS.

5.2 Flexible Software Encryption Systems

Making use of data-driven subkey selection, one can modify randomly the used elementary encryption functions and obtain different modifications of the basic conversion scheme, providing high security for each of the modifications. Reserving in the encryption subroutine some positions for simple arithmetic operations, which are to be selected from a described set (for example, $\{+_{32}, -_{32}, \oplus\}$), we can define the key-dependent modification of the set of functions e_i. In this case, the encryption procedures are not predetermined in every detail. If a large number of possible modifications ($>10^{19}$) is defined, then all users will have unique encryption algorithms defined by the secret key they select. In general, the cryptalgorithm flexibility strengthens significantly the basic conversion scheme.

Additionally, we can assign key-dependent sequence of subblocks conversion to the key-dependent selection of two-place operations. In this case the functions e_i are expressed by the formulas:

$$j(s,g') = (B^{(\alpha(g'))})^{<d(s-1)<} \bmod 2^{\delta} \tag{5.10}$$

$$B^{(\alpha(g))} \leftarrow e_i(B^{(\alpha(g))}, Q[j(s,g')]) \tag{5.11}$$

where $\alpha(g)$ is a permutation of the set of numbers $\{0, 1, 2, \ldots, w-1\}$. In general, for every ciphering subround, one can assign individual permutation, multiplying the number of the cryptalgorithm modifications by the factor $(w!)^{bR/d}$. Generation of the concrete modification is to be executed by some initialization module as a part of the precomputation procedure.

In this way, it is possible to elaborate flexible ciphers with a set of 10^{20} – 10^{200} different realizable cryptalgorithm modifications. An important problem is to evaluate the number of inequivalent modifications. For a given encryption key, any $|M|$-bit block cipher defines a permutation $\varepsilon(M)$ of the set of numbers {0, 1, 2, …, $2^{|M|} - 1$}, where $|M|$ denotes the bit length of the input data block. It is natural to use the following definition: *Two modifications of the encryption function are equivalent if they define the same permutations* $\varepsilon(M)$ *for all encryption keys.* To evaluate the number of inequivalent modifications, one has to consider the concrete cipher with flexible encryption algorithm. Because, in the proposed construction scheme, the pseudorandom subkey selection takes place for all possible modifications, we suppose that practically all different modifications are inequivalent. In the next section, we describe a flexible DDSS-based cipher with provably inequivalent cryptalgorithm modifications.

5.3 Examples of Algorithm Realization

According to different types of microprocessors used in different communication and computer systems, the following values of parameters d, δ, b, and w can be used: $d = 8, 16, 24, 32$; $\delta = 8, 16, 32$; $b = 16, 32, 64$; and $w = 2, 4, 8, 16, \ldots, 128$. In this section, an undetermined encryption algorithm oriented to 32-bit general purpose microprocessors is described. In the algorithm described in this section, four 32-bit words of the input data blocks are converted using simple arithmetic operations: bit-wise exclusive-OR operation ($\oplus$), module 2^{32} addition ($+_{32}$), module 2^{32} subtraction ($-_{32}$), and data-dependent rotations. It is supposed that a 1024-byte encryption key ($Q_0, Q_1, \ldots, Q_{255}$), represented as a sequence of 32-bit words $Q(j) = Q_j$, where $j = 0, 1, \ldots, 255$, is used. Two-place operations $\varnothing_p$ ($\varnothing_p \in \{+_{32}, -_{32}, \oplus\}$) and rotation operations $^{<c(l)<}$ defining to-left rotation by $c(l)$ bits are reserved, where $1 \le c(l) \le 31$ (p and l are the numbers of the reserved operations). The encryption key and reserved operations are to be initialized by a precomputation subroutine under the control of the secret key. The following notations are used:

- X, Y, Z, and U are 32-bit variables.
- Every variable is represented as a concatenation of 8-bit words $V = (v_1, v_2, v_3, v_4)$, where $V \in \{X, Y, Z, U\}$ and $v \in \{x, y, z, u\}$.

5.3.1 Flexible 128-Bit Cipher (d = δ = 8; b = 32; w = 4)

INPUT: 128-bit data block $M = (E, A, B, D)$.

1. Set: $R \leftarrow 3$; $m \leftarrow 1$; $U \leftarrow E$; $X \leftarrow A$; $Y \leftarrow B$; $Z \leftarrow D$.
2. $X \leftarrow [X \varnothing_{22m-21} Q(u_1)]^{<c(7m-6)<}$.
3. $Y \leftarrow [Y \varnothing_{22m-20} Q(x_1)]^{<c(7m-5)<}$.

4. $Z \leftarrow [Z \varnothing_{22m-19} Q(y_1)]^{<c(7m-4)<}$.
5. $W \leftarrow W \varnothing_{22m-18} Q(z_1)$.
6. $Z \leftarrow Z \varnothing_{22m-17} X$; $Y \leftarrow Y \varnothing_{22m-16} W$.
7. $(u_3, u_4) \leftarrow (u_3, u_4)^{<c(7m-3)<}$; $X \leftarrow X \varnothing_{22m-15} Q(u_2)$.
8. $(x_3, x_4) \leftarrow (x_3, x_4)^{<c(7m-2)<}$; $Y \leftarrow Y \varnothing_{22m-14} Q(x_2)$.
9. $(y_3, y_4) \leftarrow (y_3, y_4)^{<c(7m-1)<}$; $Z \leftarrow Z \varnothing_{22m-13} Q(y_2)$.
10. $(z_3, z_4) \leftarrow (z_3, z_4)^{<c(7m)<}$; $W \leftarrow W \varnothing_{22m-12} Q(z_2)$.
11. $Y \leftarrow Y \varnothing_{22m-11} X$; $Z \leftarrow Z \varnothing_{22m-10} U$.
12. $X \leftarrow X \varnothing_{22m-9} Q(u_3)$.
13. $Y \leftarrow Y \varnothing_{22m-8} Q(x_3)$.
14. $Z \leftarrow Z \varnothing_{22m-7} Q(y_3)$.
15. $U \leftarrow U \varnothing_{22m-6} Q(z_3)$.
16. $X \leftarrow X \varnothing_{22m-5} Q(u_4)$; $Y \leftarrow Y \varnothing_{22m-4} W$.
17. $Y \leftarrow Y \varnothing_{22m-3} Q(x_4)$; $Z \leftarrow Z \varnothing_{22m-2} X$.
18. $Z \leftarrow Z \varnothing_{22m-1} Q(y_4)$.
19. $U \leftarrow U \varnothing_{22m} Q(z_4)$.
20. If $m < R$, then increment $m \leftarrow m + 1$, and jump to step 2; otherwise STOP.

OUTPUT: 128-bit ciphertext block $C = (U, X, Y, Z)$.

One can compose decryption algorithm in accordance with equations (5.6) and (5.7). The secure number of encryption rounds is $R \geq 3$. In this flexible cipher, $22R$ binary operations and $7R$ rotation operations are reserved ($4R$ positions for rotation by 1 to 15 bits and $3R$ positions for rotation by 1 to 31 bits). The number of different realizable encryption function modifications is equal to $3^{22R} \cdot 15^{4R} \cdot 31^{3R} \approx 10^{20R}$. Using a technique proposed in studies by Moldovyan, A.A. and Moldovyan, N.A. [36], one can prove the following statement: *for $R \leq 7$, there are no two equivalent modifications of the encryption algorithm.* We can fix the arbitrary modification of the encryption algorithm and use it as an individual block cipher. A flexible 64-bit block cipher with provable inequivalent cryptalgorithm modifications is considered in the studies [36]. The cipher described corresponds to the class of ciphers introduced in Section 5.1. The next two ciphers have analogous structures, but they do not exactly correspond to that class of the DDSS-based ciphers described in Section 5.1. The difference consists in performing to-left rotation of the current controlling data subblock by eight bits, and using the eight least significant bits of the controlling subblock to perform each operation of the subkey selection. Such a modification provides simplification of the description of the DDSS-based ciphers.

5.3.2 The DDSS-Based Cipher with 64-Bit Input Data Block ($d = \delta = 8$; $b = 32$; $w = 2$)

INPUT: 128-bit data block $M = (A, B)$.

1. Set: $R \leftarrow 3$; $m \leftarrow 1$; $X \leftarrow A$; $Y \leftarrow B$.
2. $Y \leftarrow [Y +_{32} Q(x_1)]^{<8<}$.
3. $X \leftarrow X^{<8<}$.
4. $X \leftarrow X +_{32} Q(y_1)$.
5. If $m < 4R$, then increment $m \leftarrow m + 1$, and jump to step 2; otherwise STOP.

OUTPUT: 64-bit ciphertext block $C = (X, Y)$.

5.3.3 The DDSS-Based Cipher with 128-Bit Input Data Block ($d = \delta = 8$; $b = 32$; $w = 4$)

INPUT: 128-bit data block $M = (E, A, B, D)$.

1. Set: $R \leftarrow 3$; $m \leftarrow 1$; $U \leftarrow E$; $X \leftarrow A$; $Y \leftarrow B$; $Z \leftarrow D$.
2. $X \leftarrow [X +_{32} Q(u_1)]^{<8<}$.
3. $U \leftarrow U^{<8<}$.
4. $U \leftarrow U +_{32} Q(x_1)$.
5. $Z \leftarrow [Z +_{32} Q(y_1)]^{<8<}$.
6. $Y \leftarrow Y^{<8<}$.
7. $Y \leftarrow Y +_{32} Q(z_1)$.
8. $Y \leftarrow Y \oplus X$.
9. $U \leftarrow U \oplus Z$.
10. If $m < 4R$, then increment $m \leftarrow m + 1$, and jump to step 2; otherwise STOP.

OUTPUT: 128-bit ciphertext block $C = (U, X, Y, Z)$.

Decryption algorithms corresponding to the last encryption algorithms are quite evident. In the considered ciphers, the encryption and decryption processes are different, but this is not a critical point as the ciphers are oriented to software implementation.

5.4 General Characterization of the DDSS-Based Algorithms

In the ciphers already presented, four fixed operations (modulo 2^{32} addition, modulo 2^{32} subtraction, modulo 2 addition, and rotation) and one variable operation (DDSS) are used. There are other possible designs of the DDSS-based ciphers using other operations, for example, the DDR and DDP operations (if the BPI is imbedded in the microprocessor). A distinguishing feature of the described ciphers is the use of DDSS. The subkey selection is not predetermined and can be characterized as a pseudorandom one. For arbitrary modification of the flexible encryption algorithm, the transformation of some input data block M can be described by the following generalized formula:

$$C = F(M, Q_{h(1)}, Q_{h(2)}, \ldots, Q_{h(16R)}) \tag{5.12}$$

where $\{Q_{h(i)}\}$, $i = 1, 2, \ldots, 16R$, is a set of the subkeys used for encrypting the data block.

Let us estimate the security in case of a known cryptalgorithm modification. Though the sets of indices $\{h(i)\}$ are unique for all input data blocks, some repetitions of the ordered sets of subkeys can take place due to the possible presence of some equal subkeys corresponding to different indices. Solving simultaneously several equations (5.12) composed of different pairs (M, C) for which repetition takes place, one can calculate several particular sets $Q_{h(i)}$. However, the main problem consists in detecting repetitions (for attacks based on known plaintexts) or repetition generation (for attacks based on chosen texts). The security is conditioned mainly by the complexity of recognizing the $Q_{h(i)}$ set repetition, regardless of the complexity of the solution of the equations (5.12) or the system of such equations. The probability of such repetition is very low due to the low probability of having several equal subkeys in the expanded key. If the expanded key is generated pseudorandomly, then the probability that the expanded key contains at least two equal subkeys is about 2^{-16}. Besides, it is easy to design precomputation procedures such that the generated extended key contains no equal subkeys. As the repetitions of some sets $\{h(i)\}$ are not really generated, cryptanalysis can be based on solving the systems corresponding to Equation 5.12 with different sets of indices $\{h(i)\}$ but with restricted number of different subkeys considered as unknown (we have only 256 different values of subkeys for some fixed secret key). For a rough security evaluation, one can use the following probabilistic model of the 128-bit DDSS-based cipher. Let us assume the following:

1. Key elements $Q_{h(i)}$ are selected randomly.
2. To break the cipher, it is sufficient to recognize two pairs, (M_k, C_k) and (M_j, C_j), corresponding to the sets $\{Q_{h'(i)}\}$ and $\{Q_{h''(i)}\}$, respectively, for which we have $h'(i) = h''(i)$ for $i \in \{8R, 8R + 1, \ldots, 16R\}$.

The probability of the last event is $P = L^{-8R}$, where $L = 256$ (the number of subkeys). The minimum number of input blocks $N_{\min}$ that is necessary for cryptanalysis can be estimated by the formula corresponding to the 0.5 probability of the observation of such "truncated" repetition: $N_{\min} = P^{-1/2} = L^{4R} = 2^{32R}$. Let us assume that one can detect truncated repetitions by considering simultaneously just two equations similar to Equation 5.12. To detect a true repetition, it is necessary on average, to check $\binom{N_{\min}}{2}$ of different pairs (M, C). The complexity of this procedure is

$$S_{\min} = \frac{S_0}{2} \cdot \binom{N_{\min}}{2} \approx (S_0/4)\, N_{\min}^2 = (S_0/4)\, L^{8R} \tag{5.13}$$

where S_0 is the average number of operations that are to be executed to check if a repetition takes place. For $S_0 = 4$, we have $S_{min} \approx L^{8R} = 2^{64R}$ operations where $R \geq 3$. Taking into account this evaluation and the very large number of possible cryptalgorithm modifications, one can conclude that the aforementioned ciphers are secure.

5.5 Advanced DDSS-Based Ciphers

In the DDSS-based algorithms described earlier, the selected subkeys are used directly to transform a data subblock. This fact can be used in some cryptanalytic attacks, for example, in differential fault analysis (DFA), which allows one to "localize" avalanche effect and calculate independently some part of the extended key. To improve the avalanche introduced by the DDSS procedure, it is possible to use some intermediate "accumulating" variables that take on values depending on many encryption steps. The accumulation effect can be defined with the use of some simple concatenation mechanisms for calculation of the intermediate variable values. This mechanism is very effective in the ciphers with large-size input (from 128 to 4096 bits). We shall call the accumulating variables the *key variables* because they play the role of some virtual key depending on many different subkeys of the extended key. Using key variables, we avoid the direct use of subkeys in the transformation procedure of the current data subblocks. An advanced DDSS mechanism can be designed using some key variables to perform table lookup operations corresponding to DDSS. In this case, the current data subblock influences many table lookup operations, i.e., it influences many selected subkeys. Figure 5.1 illustrates the general scheme of data encryption using a cryptosystem with advanced DDSS. In the following text, we will consider possible variants realizing advanced DDSS mechanisms.

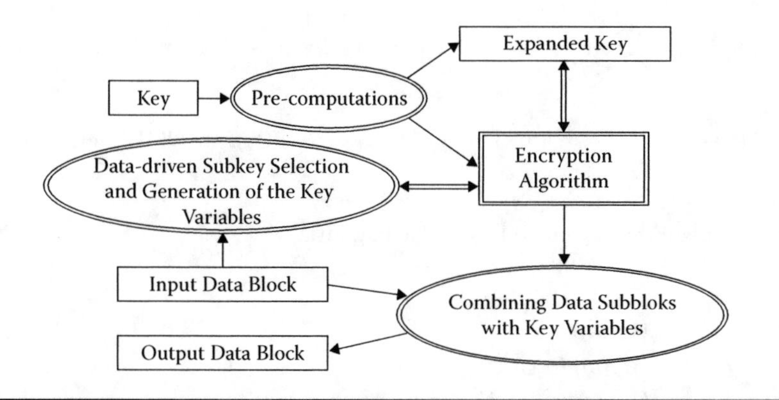

Figure 5.1 General scheme of transformations in advanced DDSS-based ciphers.

Usually, DDSS-based ciphers use some complex precomputation algorithms for generating the expanded key and initializing the encryption algorithm (in case of flexible ciphers). Different variants of the precomputation algorithms designed for the DDSS-based ciphers are considered in the literature [32–36]. The precomputation procedure is time consuming, but for many applications of the software-suitable ciphers it is not critical. Indeed, precomputations are to be performed at initial booting.

In this section, we present the variants of the advanced encryption algorithms based on DDSS. In the algorithms described in the following text, the 32-bit words of the input data blocks are converted sequentially using simple arithmetic operations, and fixed and data-dependent rotations. Security of this algorithm is based on the formation of the pseudorandom key variables U, V, and Y. The sequences of their values $\{U_i\}$, $\{V_i\}$, and $\{Y_i\}$ represent some virtual keys depending on the input message. The 32-bit elements of such virtual keys are formed on the basis of DDSS. While designing the advanced DDSS-based algorithms, the main design principle is the following: *The alteration of arbitrary input bits must change the subkey selection.* This criterion provides the generation of unique virtual keys for all input data blocks.

5.5.1 Algorithm 1: DDSS-Based Cipher with Fixed Operations

The expanded key represents a 2051-byte sequence $(q_0, q_1, \ldots, q_{2051})$ generated depending on the secret key.

INPUT: 512-byte data block represented as a sequence of 32-bit words $\{T_j\}$, $j = 0, 1, \ldots, 127$.

1. Define conversion mode: $e = 0$ (encryption) or $e = 1$ (decryption).
2. Set counter $r \leftarrow 1$ and the number of ciphering rounds $R = 4$; define $\{W_w\} = \{T_w\}$, $w = 0, 1, \ldots, 127$.
3. Set counter $i = 1$ and calculate the initial values of the variables U, V, Y, G, and n:

$$U \leftarrow Q(0),\ V \leftarrow Q(5),\ Y \leftarrow Q(11),\ G \leftarrow Q(21),\ n \leftarrow Q(31) \bmod 2^{11},$$

   where $Q(x) = (q_x, q_{x+1}, q_{x+2}, q_{x+3})$ is a subkey.
4. Calculate the current values of the key variables:
   $n \leftarrow [n \oplus (G +_{11} 0)] +_{11} U,$
   $V \leftarrow (V -_{32} G)^{>11>} \oplus Q(n),$
   $n \leftarrow n \oplus (V +_{11} 0),$
   $U \leftarrow [U -_{32} Q(n)] \oplus G^{>22>},$
   $n \leftarrow n +_{11} U,$
   $Y \leftarrow Y \oplus Q(n).$

5. Calculate the index j: $j = i - 1$, if $(-1)^{r+e} = -1$; or $j = 128 - i$, if $(-1)^{r+e} = 1$.
6. If $e = 1$, then jump to step 8.
7. Execute the current encryption step of the rth round:
   $C_j \leftarrow [(W_j -_{32} V)^{<U<}\} \oplus Y] +_{32} U$; jump to step 9.
8. Execute the current decryption step of the rth round:
   $C_w \leftarrow [(W_j -_{32} U) \oplus Y]^{>U>} +_{32} V$.
9. Update the variable G: $G \leftarrow C_j$ if $e = 0$, or $G \leftarrow W_j$ if $e = 1$.
10. If $i < 128$, then increment i, and jump to step 4.
11. If $r < R$, then increment r, define new input data block $\{W_j\} = \{C_j\}$, $j = 0, 1, \ldots, 127$, and jump to step 3; otherwise STOP.

OUTPUT: 512-byte data block $\{C_j\}$, $j = 0, 1, \ldots, 127$.

In Algorithm 1, at every elementary encryption step, three key variables V, U, and Y are used. The current values of these variables are defined by subkey selection and depend on both the encryption key and the input message. Every bit of the data blocks influences the selection of the current subkeys. The values of the variables V, U, and Y depend on pseudorandomly selected combinations of i subkeys where i is the number of the elementary ciphering step. The number (J_i) of the possible values of the variables V, U, and Y depends on i: $J_i = 1$ (for $i = 1$), $J_i \geq 2^{11(i-1)}$ (for $i = 2, 3$), and $J_i \approx 2^{32}$ (for $i = 4, 5, \ldots, 128$). Thus, for $i = 2, 3$ (for $i \geq 4$), about $2^{33(i-1)}$ ($\approx 2^{96}$) different sets $\{V, U, Y\}$ are possible. Realization of the concrete sets depends on the input message. Low probability of the appearance of the given three values $\{V, U, Y\}$ at the given step defines high level of the cryptanalysis complexity; conversion procedures of the input 32-bit words are very simple in every round, though.

5.5.2 Algorithm 2: Flexible Advanced DDSS-Based Cipher

The expanded key represents a 2051-byte sequence $\{q_0, q_1, \ldots, q_{2051}\}$ generated depending on the secret key. Two-place operations $\varnothing_p$ ($\varnothing_p \in \{+_{32}, -_{32}, \oplus\}$) and rotation operations ${}^{<u(l)<}$ defining to-left rotation by $u(l)$ bits, where $1 \leq u(l) \leq 31$ (p and l are the numbers of the reserved operations), are to be initialized, depending on the secret key.

INPUT: 512-byte data block represented as a sequence of 32-bit words $\{T_j\}$, $j = 0, 1, \ldots, 127$.

1. Set $r = 1$, $R = 4$, and define $W_w = T_w$, $w = 0, 1, 2, \ldots, 127$.
2. Set counter $i = 1$, and calculate the initial values of the variables U, V, Y, G, and n:

   $$U \leftarrow Q(1),\ V \leftarrow Q(2),\ Y \leftarrow Q(3),\ G \leftarrow Q(4),\ n \leftarrow Q(5) \bmod 2^{11},$$

   where $Q(x) = (q_x, q_{x+1}, q_{x+2}, q_{x+3})$ is a subkey.

3. Execute calculations:
   $n \leftarrow [n \oplus (G +_{11} 0)] -_{11} U$,
   $V \leftarrow (V +_{32} G)^{>11>} \varnothing_{7r-6} Q(n)$,
   $n \leftarrow n \varnothing_{7r-5} (V +_{11} 0)$,
   $U \leftarrow [U \varnothing_{7r-4} Q(n)]^{>u(3r-2)>} -_{32} (G^{>22>})$,
   $n \leftarrow n +_{11} U$,
   $Y \leftarrow Y \varnothing_{7r-3} Q(n)$.
4. Calculate the index j: $j = i - 1$, if $(-1)^r = -1$; or $j = 128 - i$, if $(-1)^r = 1$.
5. Execute the current encryption step:
   $G \leftarrow [(W_w \varnothing_{7r-2} V)^{<u(3r-1)<} \varnothing_{7r-1} Y]^{>u(3r)>}$;
   $C_w \leftarrow G \varnothing_{7r} U$.
6. Save the value C_j.
7. If $i < 128$, then increment i, and jump to step 3.
8. If $r < R$, then increment r, define $W_j = C_j$, $j = 0, 1, \ldots, 127$, and jump to step 2.
9. STOP.

OUTPUT: 512-byte data block $\{C_j\}$, $j = 0, 1, \ldots, 127$.

The number of possible modifications of the encryption Algorithm 2 is $I \approx 10^{7R}$. The decryption algorithm corresponding to this flexible cipher is quite evident. In general, the best differential and linear characteristics depend on the encryption algorithm modification. Therefore, to apply differential cryptanalysis (DCA) or linear cryptanalysis (LCA) to break a flexible cipher, a potential attacker should develop a new technique that will take into account the dependence of the best characteristics on the secret key. It is very likely that flexible ciphers with a sufficiently large number of encryption algorithm modifications are significantly more secure against DCA and LCA than the ciphers with a fixed encryption procedure. This problem is not yet studied in detail; therefore, in the following section, we roughly estimate the security of the advanced DDSS-based ciphers, assuming that the encryption algorithm modification is known.

5.6 A Model for Security Estimation

Taking into account the fact that the main feature of the advanced DDSS procedure is the pseudorandom subkey selection, the following probabilistic model for rough security estimation is proposed:

- The encryption algorithm modification is known.
- To break the cipher means finding a simultaneous repetition of the values of key variables U, V, and Y that are used while converting some input 32-bit words (data subblocks).

- The work effort for detection of one repetition defines the security level of the cipher.
- In the rth round, where $r \geq 3$, the key variables change at random.

For example, in Algorithms 1 and 2 it is supposed that some input word T_w was converted with six 32-bit random virtual subkeys $\{U_3, V_3, Y_3, U_4, V_4\ Y_4\}_w$ where indices 3 and 4 denote the number of the encryption round. The current set $\{U_3, \ldots, Y_4\}_w$ is equal to the set $\{U_3, \ldots, Y_4\}_0$ corresponding to some word T_0 with the probability $P = J^{-6}$, where $J = 2^{32}$. The value $N \approx P^{-1/2} = J^3$ is the number of the input words T, for which repetition takes place with probability 0.5. This value is the evaluation of the minimal size of the encrypted text $v_{\min}$, which is necessary for cryptanalysis. To detect a repetition with high probability, an attacker has to test about $N^2/2$ couples of input words, and the work effort is

$$S_{\min} \approx s_r J^6/2 \geq 2^{191}\ (operations) \tag{5.14}$$

where s_r is the complexity of some repetition recognition criteria in some standard operations ($s_r \geq 1$ operation).

The advanced DDSS-based ciphers are resistant to attacks on the basis of the known part Δ of the expanded key. We call such attacks *known-key attacks*. If the known part of the expanded key is sufficiently large, then an attacker is able to select such input data block that, in the first encryption round, only known subkeys will be selected. However, predetermination of the subkey selection in the second and third rounds is a difficult problem. To consider a detailed known-key attack, one should consider a concrete encryption algorithm. However, some generalized estimations can be obtained in the frame of the probabilistic model.

Let us assume that the attacker is able to choose such input data block that, in the first round, only known subkeys are selected (i.e., subkeys corresponding to the known part of the expanded key). This is possible because the input words T define the selection of the subkeys. In rounds with number $r \geq 2$, the subkey selection is pseudorandom, except several initial elementary encryption steps. If known part Δ is large (for example, $\Delta = 0.997$), then there is a high probability that, in all rounds, only known subkeys will be selected. In this case, it is easy to calculate unknown subkeys (their values and locations) selected in the final round. At every elementary encryption step, three subkeys are selected. The probability of their selection from the known part of the encryption key is $P_1 = \Delta^3$ for each elementary encryption step of rounds $r = 2, 3, \ldots, R - 1$. The probability to select only known subkeys in $R - 2$ rounds is

$$P_2 = \Delta^{3E\,(R-2)} \tag{5.15}$$

where E is the total number of encryption steps in one round (for the 512-byte input data block, we have $E = 128$). A DDSS-based cipher is secure against known-key

Table 5.1 Rough security estimation results

Cipher	I[a]	I_{min}	V_{min}, *bytes*	S_{min}, *operations*	Δ_c
Algorithm 1 ($R = 3$)	1	1	10^{15}	10^{28}	0.82
Algorithm 1 ($R = 4$)	1	1	10^{29}	10^{57}	0.91
Algorithm 2 ($R = 4$)	10^{28}	$>10^{17}$	10^{29}	$>10^{73}$	0.91[b]

[a] Number of different modifications of the encryption algorithm.
[b] In case of known cryptalgorithm modification.

attack if the probability P_2 is sufficiently small, i.e., $P_2 \le P_a$, where P_a is a sufficiently small value (for example, $P_a < 10^{-20}$). The critical value of the known part of the expanded key Δ_c corresponds to the condition $P_2 \le P_a$. For $E \ge 120$, from Equation 5.3 we can obtain

$$\Delta_c \ge \exp\frac{\ln P_a}{360(R-2)} \tag{5.16}$$

where R is the assigned number of the rounds.

Taking into account the fact that the elementary operations in Algorithm 2 are not known, the expected resultant security of this algorithm can be evaluated by the formula

$$S_{min} = S'_{min}(I_{min}/2) \tag{5.17}$$

where S'_{min} is the estimation corresponding to the case of known key-dependent operations, and I_{min} ($I_{min} \approx 2.4^R \cdot 10^{4R} > 10^{17}$) is the number of possible sets of operations used for conversion of individual 32-bit words. The parameters characterizing the security of Algorithms 1 and 2 are presented in Table 5.1, where the value Δ_c corresponds to the case $P_a = 10^{-30}$.

5.7 A DDSS-Based Cipher with Flexible Input Data Block Size

Software encryption algorithm Spectr-F considered in this section represents another design of DDSS-based algorithms. The letter F denotes that the size of the input block is flexible and can be set from 128 bits to $32Z$ bits, where $Z > 4$. Due to this, Spectr-F has the necessary flexibility for optimization of the block length choice in concrete applications. It provides some benefits in speed when encrypting large data blocks (Spectr-F has performance of 1000 Mbit/s and more for the general purpose processors available today). Spectr-F includes two full rounds and four shortcut rounds that make it significantly faster than Algorithms 1 and 2 in case of $Z \geq 16$. Shortcut rounds are used to improve the avalanche effect corresponding to the last 32-bit words of the input data block.

5.7.1 Design Criteria

The Spectr-F cipher is intended for a wide range of applications, including encryption in fast telecommunication channels and computer systems, namely, maintenance of the real-time automatic encryption of the data stored in long-term storage devices. When using processors such as Pentium, it is necessary to provide encryption speeds ranging from hundreds up to a thousand Mbps.

The variable length of the input block is applied to provide more flexibility to the Spectr-F algorithm. A parameterized value $32Z$ (bits), where Z is a natural number and $Z \geq 4$, was chosen as the size of the input data block of Spectr-F algorithm. When designing this algorithm, the following criteria are used:

1. The encryption routing of the input data block must be carried out as the consecutive transformation of 128 words, each word having the length of 32 bits. It should only use operations that are fast in software: two-place operations $+_{32}$, $-_{32}$, $\oplus$; rotations $^{>>>}$, $^{<<<}$; operation of the exchange of values between data registers; and table lookup operations.
2. Each bit of all words transformed in the previous encryption steps must have an essential influence on the transformation of all consecutive words, i.e., the consecutive transformation of words must be carried out in concatenation mode. It will provide a strong avalanche effect when passing from the first to the last words of the input block. As it is supposed to execute many elementary transformation steps (by which the 32-bit data subblocks are transformed) in one round, this design criterion can provide high security, although assigning few encryption rounds.

To set an effective concatenation mode, it is necessary to use at least two variables, the current values of which will be calculated based on their earlier values and the value of the current transformed word. We have selected the DDSS as the basic mechanism of transformation because it has sufficient approbation [32–37]. The DDSS appears to be a much more attractive cryptographic primitive in comparison with substitution operations performed with the use of some key-dependent tables.

5.7.2 Local Notations

Throughout the next subsections, we shall use the following operations, symbols, and notation:

- The term "word" is used to denote a 32-bit data subblock.
- We shall denote an element of the sequence L by L_i or $L[i]$, where $i = 0, 1, 2, \ldots$.
- We shall also interpret the sequence of bytes $l_0, l_1, \ldots, l_n$ as a sequence of the 32-bit words $L_0, L_1, \ldots, L_s$, where $L_j = \{l_{4j}, l_{4j+1}, l_{4j+2}, l_{4j+3}\}$ and $j = 0, 1, \ldots, s$; while interpreting several sequential bytes by a binary number, the right byte corresponds to the most significant digits.
- "$>>>$" denotes right circular rotation of words; the cyclic rotation of word X to the right by Y bits is denoted "$X^{>>>Y}$" (note that in such data-dependent rotations only $\log_2 32 = 5$ least significant bits of Y are used to determine the rotation amount).
- "$\leftrightarrow$" denotes exchange operation; for example, $W \leftrightarrow V$ produces the same result as the following three operations: $T \leftarrow W$, $W \leftarrow V$, and $V \leftarrow T$.
- Constant F is equal to FFFF07FF, i.e., $F = (1,1,1,1,1,1,1,1,1,1,1,1,0,0,0,0,0,1,1, \ldots, 1)$.

5.7.3 Transformation Algorithms

The Spectr-F cipher uses a 2051-byte expanded key $(q_0, q_1, \ldots, q_{2051})$ that is to be generated depending on the secret key. While data ciphering, the subkeys $Q_x = (q_x, q_{x+1}, q_{x+2}, q_{x+3})$ are selected based on some controlling data subblock. The main requirements of the expanded key generation procedure are: (1) uniform influence of each bit of the secret key on all bits of the generated expanded key and (2) high work effort for the calculation of the user's secret key using a known extended key. The Spectr-F encryption algorithm includes two full and four reduced encryption rounds. Plaintext block T is divided into 32-bit words T_i: $T = (T_0, T_1, \ldots, T_{Z-1})$, where $Z \geq 4$. The value of the natural number Z is determined depending on the area of application. In each round, the consecutive transformation of input 32-bit words $T_0, T_1, \ldots, T_{Z-1}$ is carried out; in case $Z = 4$, the complete and reduced rounds are identical. After each round, except for the last one, the exchange of values in the following word pairs is carried out: $T_0 \leftrightarrow T_3$ и $T_1 \leftrightarrow T_2$. The transformation algorithms include the following two typical procedures: ***Initialize*** and ***Change_NRVYU***.

Procedure ***Initialize***

1. Initialize the value of the internal counter $i \leftarrow 0$ and the number $n \leftarrow Q_2 +_{11} 0$.
2. Set initial values of the variables $R \leftarrow Q_9$, $V \leftarrow Q_7$, $Y \leftarrow Q_3$, $U \leftarrow Q_9$, $N \leftarrow Q_5$.
3. END.

Procedure ***Change_NRVYU***

1. $N \leftarrow N \oplus R$; $V \leftarrow V +_{32} N$;
2. $N \leftarrow N \otimes F$; $n \leftarrow N +_{11} 0$; $V \leftarrow (V +_{32} Q_n)^{<<<11}$;

3. $N \leftarrow N \oplus V$; $Y \leftarrow Y +_{32} N$;
4. $N \leftarrow N \otimes F$; $n \leftarrow N +_{11} 0$; $Y \leftarrow (Y +_{32} Q_n)^{<<<11}$;
5. $N \leftarrow N +_{32} Y$; $N \leftarrow N \otimes F$; $n \leftarrow N +_{11} 0$;
6. $U \leftarrow ((U \oplus Q_n) +_{32} R)^{<<<V}$.
7. END.

Spectr-F cipher is described by the following algorithms.

Encryption procedure in four reduced rounds

1. Initialize the value of the external counter $j \leftarrow 0$.
2. Execute ***Initialize*** procedure.
3. Execute ***Change_NRVYU*** procedure.
4. Transform the current word T_i: $T_i \leftarrow (T_i -_{32} V) \oplus U$.
5. Convert the variable R: $R \leftarrow R +_{32} T_i$.
6. Finish transformation of the current word T_i: $T_i \leftarrow T_i^{<<<V} -_{32} Y$.
7. Increment $i \leftarrow i + 1$. If $i \neq 4$, then go to step 3.
8. Perform the exchange of values in the following word pairs: $T_0 \leftrightarrow T_3$ and $T_1 \leftrightarrow T_2$.
9. Increment $j \leftarrow j + 1$. If $j \neq 4$, then go to step 2.
10. STOP.

Decryption procedure in four reduced rounds

1. Initialize the value of the external counter $j \leftarrow 0$.
2. Execute procedure ***Initialize***.
3. Execute procedure ***Change_NRVYU***.
4. Transform the current word T_i: $T_i \leftarrow (T_i +_{32} Y)^{>>>V}$.
5. Convert the variable R: $R \leftarrow R +_{32} T_i$.
6. Finish transformation of the current word T_i : $T_i \leftarrow (T_i \oplus U) +_{32} V$.
7. Increment $i \leftarrow i + 1$. If $i \neq 4$, then go to step 3.
8. Perform the exchange of values in the following word pairs: $T_0 \leftrightarrow T_3$ and $T_1 \leftrightarrow T_2$.
9. Increment $j \leftarrow j + 1$. If $j \neq 4$, then go to step 2.
10. STOP.

The first (full) encryption round

1. Execute procedure ***Initialize***.
2. Execute procedure ***Change_NRVYU***.
3. Transform the current word $T_i \leftarrow (T_i -_{32} V) \oplus U$.
4. Convert the variable R: $R \leftarrow R +_{32} T_i$.
5. Finish transformation of the current word T_i: $T_i \leftarrow T_i^{<<<V} -_{32} Y$.
6. Increment $i \leftarrow i + 1$. If $i \neq Z$, then go to step 2.
7. Transform the words T_2 and T_3 as follows: $T_2 \leftarrow T_2 \oplus T_{Z-2}$ and $T_3 \leftarrow T_3 \oplus T_{Z-1}$.

8. Exchange the values of the words in pairs (T_0,T_3) and (T_1,T_2): $T_0 \leftrightarrow T_3$ and $T_1 \leftrightarrow T_2$.
9. STOP.

The procedure of the sixth (full) encryption round

1. Transform the words T_3 and T_2 as follows: $T_3 \leftarrow T_3 \oplus T_{Z-1}$ and $T_2 \leftarrow T_2 \oplus T_{Z-2}$.
2. Execute procedure ***Initialize.***
3. Execute procedure ***Change_NRVYU.***
4. Transform the current word $T_i \leftarrow (T_i -_{32} V) \oplus U$.
5. Convert the variable R: $R \leftarrow R +_{32} T_i$.
6. Finish transformation of the current word T_i: $T_i \leftarrow T_i^{<<<V} -_{32} Y$.
7. Increment $i \leftarrow i + 1$. If $i \neq Z$, then go to step 3.
8. STOP.

The first (full) decryption round

1. Execute procedure ***Initialize.***
2. Execute procedure ***Change_NRVYU.***
3. Transform the current word $T_i \leftarrow (T_i +_{32} Y)^{>>>V}$.
4. Convert the variable R: $R \leftarrow R +_{32} T_i$.
5. Finish transformation of the current word T_i: $T_i \leftarrow (T_i \oplus U) +_{32} V$.
6. Increment $i \leftarrow i + 1$. If $i \neq Z$, then go to step 2.
7. Transform the words T_2 and T_3 as follows: $T_2 \leftarrow T_2 \oplus T_{Z-2}$ and $T_3 \leftarrow T_3 \oplus T_{Z-1}$.
8. Exchange the values of the words in pairs (T_0,T_3) and (T_1,T_2): $T_0 \leftrightarrow T_3$ and $T_1 \leftrightarrow T_2$.
9. STOP.

The procedure of the sixth (full) decryption round

1. Transform the words T_3 and T_2 as follows: $T_3 \leftarrow T_3 \oplus T_{Z-1}$ and $T_2 \leftarrow T_2 \oplus T_{Z-2}$.
2. Execute procedure ***Initialize.***
3. Execute procedure ***Change_NRVYU.***
4. Transform the current word $T_i \leftarrow (T_i +_{32} Y)^{>>>V}$.
5. Convert the variable R: $R \leftarrow R +_{32} T_i$.
6. Finish transformation of the current word T_i: $T_i \leftarrow (T_i \oplus U) +_{32} V$.
7. Increment $i \leftarrow i + 1$. If $i \neq Z$, then go to step 3.
8. STOP.

The scheme of the plaintext transformation is shown in Figure 5.2.

The security of Spectr-F has been estimated using the model described in Section 5.6. The results are presented in Table 5.2.

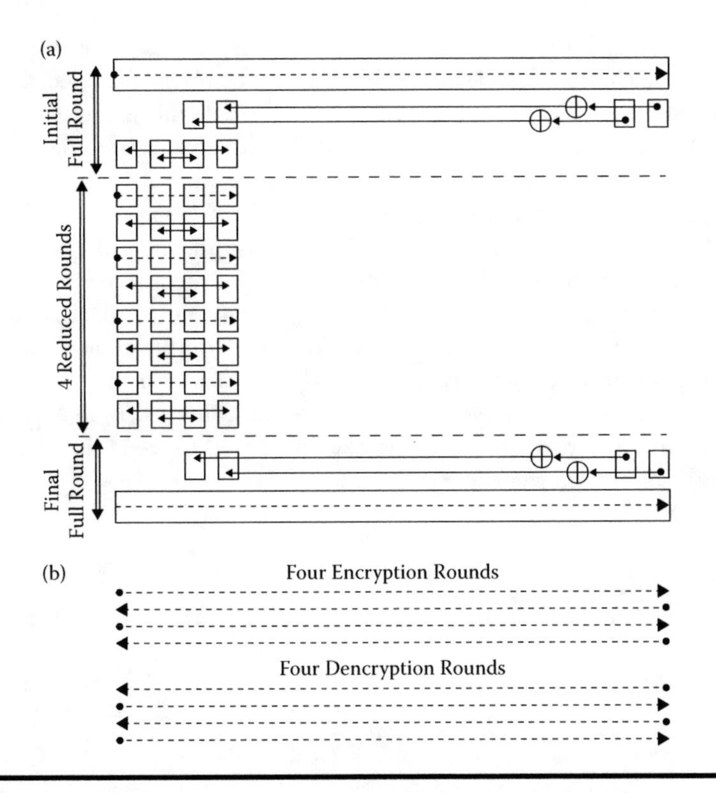

Figure 5.2 Scheme of the consecutive transformation of the 32-bit words in the advanced DDSS-based ciphers: (*a*) Spectr-F; (*b*) Algorithms 1 and 2.

Table 5.2 Rough security estimation results for Spectr-F

Type of the attack	*P*	Size of minimum required text, # *words*	*S*, # *operations*
Known plaintext	2^{-192}	2^{96}	2^{191}
Chosen text	2^{-96}	2^{48}	2^{95}

5.8 Conclusions

The DDSS is an efficient primitive for designing fast software-suitable ciphers. To implement such ciphers, sufficiently large expanded keys should be used, which are to be precomputed depending on the secret key. The advanced DDSS-based ciphers are secure against known-plaintext, chosen-text, and known-key attacks. The proposed flexible ciphers are secure against these attacks in cases in which the encryption algorithm modification is known. DFA attack is efficient against the ciphers (including flexible ones) that directly combine the selected subkeys with data subblocks. However, advanced DDSS-based ciphers are secure against such DFA attacks. More security can be added into fast software-oriented ciphers based on the DDSS by using the mechanism of generating encryption algorithms depending on the secret key. Fast flexible-block DDSS-based ciphers with a large number of possible modifications of the encryption function (10^{7R}, where $R \geq 2$) are presented. The use of the key-dependent encryption algorithms strengthens the fast software encryption systems against the majority of known attacks. The ciphers with flexible algorithms can be tested using known-algorithm attacks. It is reasonable to test flexible algorithms using chosen-algorithm attacks, which presupposes that an attacker is able to select the arbitrary possible modification of the encryption algorithm. In other words, the attacker is allowed to select the modification that has the minimum security.

References

1. Advanced Encryption Standard. 1997. Proceedings of the 4th International Workshop, Fast Software Encryption—FSE '97, Ed. Eli Biham. *Springer Verlag LNCS* 1267: 83–87.
2. Becker, W. 1979. Method and system for machine enciphering and deciphering. U.S. Patent # 4157454.
3. Benes, V.E. 1962. Algebraic and topological properties of connecting networks. *Bell Systems Technical Journal* 41: 1249–74.
4. Benes, V.E. 1965. *Mathematical theory of connecting networks and telephone traffic.* New York: Academic Press.
5. Biham Eli and Adi Shamir. 1993. *Differential cryptanalysis of the data encryption standard.* Berlin: Springer-Verlag, New York: Heidelberg.
6. Biryukov, A. 1999. Methods of cryptanalysis. Research thesis, Israel institute of Technology, Haifa.
7. Biryukov, A. and Wagner, D. 2000. Advanced Slide Attacks. *Springer Verlag LNCS* 1807: 589–606.
8. Bodrov, A.V., Moldovyan, A.A., and Moldovyanu, P.A. 2005. DDP-based ciphers: Differential analysis of SPECTR-H64. *Computer Science Journal of Moldova* 13: 268–91.
9. Cheung, O.Y.H., Tsoi, K.H., Leong, P.H.W., and Leong, M.P. 2001. Tradeoffs in parallel and serial implementations of the international data encryption algorithm. *Springer-Verlag LNCS* 2162: 333–37.
10. Chitu, C. and Glesner, M. 2005. An FPGA Implementation of the AES-Rijndael in OCB/ECB modes of operation. *Microelectronics Journal, Elsevier Science* 36: 139–46.
11. Clos, C. 1953. A study of nonblocking switching networks. *Bell System Technical Journal* 32: 406–24.
12. Elbirt, A.J., Yip, W., Ghetwynd, B., and Paar, C. 2000. FPGA implementation and performance evaluation of the AES block cipher candidate algorithm finalists. Proceedings of the 3rd Advanced Encryption Standard Conference. New York.
13. Goots, N.D, Izotov, B.V, Moldovyan, A.A., and Moldovyan, N.A. 2003. Fast ciphers for cheap hardware: differential analysis of SPECTR-H64. *Springer-Verlag LNCS* 2776: 449–52.

14. Goots, N.D, Moldovyan, A.A., and Moldovyan, N.A. 2001. Fast encryption algorithm SPECTR-H64. *Springer-Verlag LNCS* 2052: 275–86.
15. Ichikawa, T., Kasuya, T., and Matsui, M. 2000. Hardware evaluation of the AES finalists. Proceedings of the 3rd Advanced Encryption Standard Conference. New York. (http://www.nist.gov/aes).
16. Izotov, B.V., Moldovyan, A.A., and Moldovyan, N.A. 2001. *Controlled operations as a cryptographic primitive. Springer-Verlag LNCS* 2052: 230–41.
17. Izotov, B.V. 2003. Cryptographic primitives based on cellular transformations. *Computer Science Journal of Moldova* 11: 269–91.
18. Kam, J.B. and Davida, G.I. 1979. Structured design of substitution-permutation encryption networks. *IEEE Transactions on Computers* 28: 747–53.
19. Kitsos, P., Sklavos, P., and Koufopavlou, O. 2002. Hardware implementation of the SAFER + encryption algorithm for the bluetooth system. Proceedings of IEEE International Symposium on Circuits & Systems (ISCAS'02). U.S. Vol. IV, pp. 878–81.
20. Kitsos, P., Sklavos, P., Papadomanolakis, K., and Koufopavlou, O. 2003. Hardware implementation of the bluetooth security. *IEEE Pervasive Computing, Mobile and Ubiquitous Systems* 2: 21–29.
21. Ko, Y., Hong, S., Lee, W., Lee, S., and Kang, J.-S. 2004. Related Key differential attacks on 27 round of XTE and full-round GOST. *Springer-Verlag LNCS* 3017: 299–316.
22. Ko, Y., Hong, D., Hong, S., Lee, S., and Lim, J. 2003. Linear cryptanalysis on SPECTR-H64 with higher order differential property. *Springer-Verlag LNCS* 2776: 298–307.
23. Kwan, M. 1997. The design of the ICE encryption algorithm. *Springer-Verlag LNCS* 1267: 69–82.
24. Lee, C., Hong, D., Lee, S., Yang, H., and Lim, J. 2002. A chosen plaintext linear attack on block cipher CIKS-1. *Springer-Verlag LNCS* 2513: 456–68.
25. Lee, R.B., Shi, Z.J., and X. Yang. 2001. Efficient permutation instructions for fast software cryptography. *IEEE Micro* 21: 56–69.
26. Lee, R.B., Shi, Z.J., Rivest, R.L., and Robshaw, M.J.B. 2004. On permutation operations in cipher design. Proceedings of the International Conference on Information Technology: Coding and Computing (ITCC'04*)*. Las Vegas. Vol. 2, pp. 569–79.
27. Madryga, W.E. 1984. A high performance encryption algorithm. In *Computer security: A global challenge*, 557–70. Elsevier Science Publishers.
28. Matsui, M. 1994. Linear cryptoanalysis method for DES cipher. *Springer Verlag LNCS* 765: 386–97.
29. McLoone, M. and McCanny, J.V. 2001. High performance single-chip FPGA Rijndael algorithm implementation. *Springer-Verlag LNCS* 2162: 65–76.
30. Menezes, A.J., van Oorschot, P.C., and Vanstone, S.A. 1996. *Handbook of Applied Cryptography*. CRC Press, Boca Raton, FL.
31. Moldovyan, A.A. 2000. Fast block ciphers based on controlled permutations. *Computer Science Journal of Moldova* 8: 270–83.
32. Moldovyan, A.A., Moldovyan, N.A., and Moldovyanu, P.A. 1994. Effective software-oriented cryptosystem in complex PC security software. *Computer Science Journal of Moldova* 2: 269–82.

33. Moldovyan, A.A. and Moldovyan, N.A. 1995. Fast software encryption systems for secure and private communication. Proceedings of the 12th International Conference on Computer Communication. Seoul. Vol. 1, pp. 415–20.
34. Moldovyan, A.A., Moldovyan, N.A., and Ya. Sovetov, B. 1997. Software-oriented ciphers for computer communication protection. Proceedings of the International Conference "Applications of Computer Systems"—ACS'97. Szczecin, Poland. pp. 443–50.
35. Moldovyan, A.A., Moldovyan, N.A., and Zaikin, O.A. 1997. Undetermined software-oriented ciphers based on data-dependent subkey selection. Proceedings of the International Conference "Computer Methods in Control Systems"—CMCS'97. Szczecin, Poland. pp. 157–64.
36. Moldovyan, A.A. and Moldovyan, N.A. 1998. Flexible block ciphers with provably inequivalent cryptalgorithm modifications. *Cryptologia* XXII: 134–40.
37. Moldovyan, A.A. and Moldovyan, N.A. 1998. Software encryption algorithms for transparent protection technology. *Cryptologia* XXII: 56–68.
38. Moldovyan, A.A. and Moldovyan, N.A. 2002. A cipher based on data-dependent permutations. *Journal of Cryptology* 15: 61–72.
39. Moldovyan, A.A., Moldovyan, N.A., and Sklavos, N. 2004. Minimum size primitives for efficient VLSI implementation of DDO-based ciphers. Proceedings of the MELECON 2004. Dubrovnik, Croat. pp. 807–10.
40. Moldovyan, A.A., Moldovyan, N.A., and Sklavos, N. 2006. Controlled elements for designing ciphers suitable to efficient VLSI implementation. *Telecommunication Systems* 32: 149–63.
41. Moldovyan, N.A. 2003. On cipher design based on switchable controlled operations. *Springer-Verlag LNCS* 2776: 316–27.
42. Moldovyan, N.A. 2003. Fast DDP-based ciphers: Design and differential analysis of Cobra-H64. *Computer Science Journal of Moldova* 11: 292–315.
43. Moldovyan, N.A., Eremeev, M.A., Sklavos, N., and Koufopavlou, O. 2004. New class of the FPGA efficient cryptographic primitives. Proceedings of the ISCAS 2004. Vancouver, Canada. Vol. II, pp. 553–56.
44. Moldovyan, N.A., Moldovyan, A.A., Eremeev, M.A., and Summerville, D.H. 2004. Wireless networks security and cipher design based on data-dependent operations: Classification of the FPGA suitable controlled elements. Proceedings of the CCCT-2004. Vol. VII, Austin, TX. pp. 123–28.
45. Moldovyan, N.A., Moldovyan, A.A., and Eremeev, M.A. 2006. A class of data-dependent operations. *International Journal of Network Security* 2: 187–204.
46. Moldovyan, N.A., Moldovyan, A.A., Eremeev, M.A., and N. Sklavos. 2006. New class of cryptographic primitives and cipher design for network security. *International Journal of Network Security* 2: 114–25.
47. Moldovyan, N.A., Moldovyan, A.A., and Goots, N.D. 2005. Variable bit permutations: Linear characteristics and pure VBP-based cipher. *Computer Science Journal of Moldova* 13: 84–109.
48. Moldovyan, N.A., Moldovyanu, P.A., and Summerville, D.H. 2007. On software implementation of fast DDP-based ciphers. *International Journal of Network Security* 4: 81–89.

49. Moldovyan, N.A., Sklavos, N., and Koufopavlou, O. 2005. Pure DDP-base cipher: Architecture analysis, Hardware Implementation cost and Performance up to 6.5 Gbps. *International Arab Journal of Information Technology* 2: 24–32.
50. Moldovyan, N.A., Sklavos, N., Moldovyan, A.A., and Koufopavlou, O. 2005. Chess-64, A block cipher based on data-dependent operations: Design variants and hardware implementation efficiency. *Asian Journal of Information Technology* 4: 320–28.
51. Parker, S. 1980. Notes on shuffle/exchange-type switching networks. *IEEE Transactions on Computers* C-29: 213–22.
52. Pieprzyk Josef, Hardjono Thomas, and Jennifer Seberry. 2003. *Fundamentals of Computer Security*. Springer-Verlag, Berlin.
53. Portz, M. 1992. A Generalized Description of DES-based and Benes-based permutation generators. *Springer-Verlag LNCS* 718: 397–409.
54. Preneel, B., Bosselaers, A., and Rijmen, V. et al. 2000. Comments by the NESSIE Project on the AES Finalists. http://www.nist.gav/aes.
55. Preneel, B. et al. 2003. Performance of Optimized Implementations of the NESSIE Primitives. Project IST-1999–12324, p. 117. https://www.cosic.esat.kuleuven.be/nessie/.
56. Rivest, R.L. 1995. The RC5 encryption algorithm. *Springer-Verlag LNCS* 1008: 86–96.
57. Rivest, R.L., Robshaw, M.J.B., Sidney, R., and Yin, Y.L. 1998. The RC6 block cipher. In Proceedings of the 1st Advanced Encryption Standard Candidate Conference. Venture, California.
58. Rompay, V.B., Knudsen, L.R., and Rijmen, V. 1998. Differential cryptanalysis of the ICE encryption algorithm. *Springer-Verlag LNCS* 1372: 270–83.
59. Rudra, A., Dubey, P.K., and Jutla C.S. et al. 2001. Efficient Rijndael encryption implementation with composite field arithmetic. *Springer-Verlag LCNS* 2162: 171–80.
60. Schneier, Bruce. 1996. *Applied cryptography—protocols: Algorithms and source code in C*. Second edition. New York: John Wiley and Sons.
61. Schubert, A. and Anheier, W. 1999. Efficient VLSI implementation of modern symmetric block ciphers. Proceedings of The 6th IEEE International conference on Electronics, Circuits and Systems ICECS'99. Pafos, Cyprus.
62. Seberry, J., Zhang, X-M., and Zheng, Y. 1994. Nonlinearly balanced Boolean functions and their propagation characteristics. *Springer-Verlag LNCS* 773: 49–60.
63. Sklavos, N. and Koufopavlou, O. 2002. Architectures and VLSI implementations of the AES-proposal Rijndael. *IEEE Transactions on Computers* 51: 1454–59.
64. Sklavos, N., Koufopavlou, O., and Moldovyan, A.A. 2003. Encryption and data dependent permutations: Implementation cost and performance evaluation. *Springer-Verlag LNCS* 2776: 337–48.
65. Sklavos, N., Moldovyan, N.A., and Koufopavlou, O. 2005. High speed networking security: Design and implementation of two new DDP-based ciphers. *Mobile Networks and Applications* 10: 219–31.
66. Sklavos, N. and Koufopavlou, O. 2003. Architectures and FPGA Implementations of the SCO (-1, -2, -3) Ciphers Family. Proceedings of the 12th International Conference on Very Large Scale Integration, (IFIP VLSI SOC '03). Darmstadt, Germany. pp. 68–73.
67. Shannon, C.E. 1949. Communication theory of secrecy systems. *Bell Systems Technical Journal* 28: 656–715.

68. Waksman, A.A. 1968. Permutation network. *Journal of the ACM* 15: 159–63.
69. Weeks B., Bean M., Rozylowicz T., and Ficke, C. 2000. Hardware performance simulations of round 2 advanced encryption standard algorithms. Proceedings of the 3rd Advanced Encryption Standard (AES) Candidate Conference. New York.
70. Wilcox, D.C., Pierson, L.G., and Robertson, P.J. et al. 1999. A DES ASIC suitable for network encryption at 10 Gbps and beyond. *Springer-Verlag LNCS* 1717: 37–48.
71. Zimmermann, R., Curiger, A., and Bonnenberg, H. et al. 1994. A 177 Mb/s VLSI implementation of the international data encryption algorithm. *IEEE Journal of Solid State Circuits*, Vol. 29, No 3.

Index